Wide Area 2D/3D Imaging

Benjamin Langmann

Wide Area 2D/3D Imaging

Development, Analysis and Applications

Springer Vieweg

Benjamin Langmann
Hannover, Germany

Also PhD Thesis, University of Siegen, 2013

ISBN 978-3-658-06456-3   ISBN 978-3-658-06457-0 (eBook)
DOI 10.1007/978-3-658-06457-0

The Deutsche Nationalbibliothek lists this publication in the Deutsche Nationalbibliografie; detailed bibliographic data are available in the Internet at http://dnb.d-nb.de.

Library of Congress Control Number: 2014942840

Springer Vieweg
© Springer Fachmedien Wiesbaden 2014

Printed on acid-free paper

Springer Vieweg is a brand of Springer DE.
Springer DE is part of Springer Science+Business Media.
www.springer-vieweg.de

Acknowledgements

Conducting this work was made possible in the helpful and fruitful environment of the Center for Sensor System (ZESS) at the University of Siegen. I would like to thank the laboratory staff of the ZESS for the practical help and especially my supervisors Dr.-Ing. Klaus Hartmann and Prof. Dr.-Ing. Otmar Loffeld as well as Dr. Wolfgang Weihs for the valuable discussions and their support. Special thanks go to my second examiner Prof. Dr.-Ing. Andreas Kolb for his encouragement as well as the scientific guidance. Moreover, I would like to thank my sister Mirjam Spies for proof reading and my parents for their continued support during my studies.

Benjamin Langmann

Abstract

Imaging technology is widely utilized in a growing number of disciplines ranging from gaming, robotics and automation to medicine. In the last decade also 3D imaging found increasing acceptance and application, which were largely driven by the development of novel 3D cameras and measuring devices. These cameras are usually limited to indoor scenes with relatively low distances. In this thesis the development and the evaluation of medium and long-range 3D cameras are described in order to overcome these limitations. The MultiCam, a monocular 2D/3D camera which incorporates a color and a depth imaging chip, forms the basis for this research. The camera operates on the Time-of-Flight (ToF) principle by emitting modulated infrared light and measuring the round-trip time. In order to apply this kind of camera to larger scenes, novel lighting devices are required and will be presented in the course of this work. On the software side methods for scene observation working with 2D and 3D data are introduced and adapted to large scenes. An extended method for foreground segmentation illustrating the advantages of additional 3D data is presented, but also its limitations due to the lower resolution of the depth maps are addressed.

Long-range depth measurements with large focal lengths and 3D imaging on mobile platforms are easily impaired by involuntary camera motions. Therefore, an approach for motion compensation with joint super-resolution is introduced to facilitate ToF imaging in these areas. The camera motion is estimated based on the high resolution color images of the MultiCam and can be interpolated for each phase image, i.e. raw image of the 3D imaging chip. This method was applied successfully under different circumstances.

A framework for multi-modal segmentation and joint super-resolution also addresses the lower resolution of the 3D imaging chip. It resolves the resolution mismatch by estimating high resolution depth maps while performing the segmentation. Subsequently, a global multi-modal and multi-view tracking approach is described, which is able to take advantage of any type and number of cameras. Objects are modeled with ellipsoids and their appearance is modeled with color histograms as well as density estimates. The thesis concludes with remarks on future developments and the application of depth cameras in new environments.

Contents

List of Figures

1 Introduction

In a world with an increasing amount and availability of computers or micro-processors, the interaction between the real world and computers becomes more significant. Optical sensors play an important role in this interaction, but cameras, i.e. optical sensor matrices, have become more prominent since they have dropped in price in recent years. Despite posing substantially higher demands on data transmission and processing, many consumer devices from cell-phones to cars are today equipped with at least one camera. In the field of gaming cameras have become a standard input device and in medical engineering and science cameras also find more and more application. Moreover, thanks to the advancement of processing power cameras are standard sensors in industrial automation and robotics. Since many image processing algorithms can be implemented on DSPs or FPGAs, high performance CPUs are not required for all applications.

Nevertheless, the limitation of the available processing power was a key obstacle, which drove the development of depth imaging sensors in the past decade. Restrictions of stereo camera setups in respect of reliability and precision proved also to be problematic. In general, two different approaches were followed, namely the structured light and the time-of-flight approach. The structured or coded light principle is based on the observation of disparities similar to the stereo camera method. However, the disparities are observed between a camera and a light projector instead of two cameras. On the one hand this approach aims at overcoming the ambiguities encountered in stereo setups, e.g. untextured objects, and on the other hand it aims at reducing the computational demand. Structured light approaches were dominated for a long time by line patterns projected in a sequence with increasing spatial resolution. However, recently advances were made with highly resolved dot patterns, which significantly impacted the gaming market in addition to research and development.

The rival distance measuring principle is called Time-of-Flight (ToF) and here the distance of an object to the camera is determined by emitting spatially uniform light, which is varied over time, thus allowing to measure the time until the light is received. Cameras following this operating principle have been researched for more than a decade and models from several manufacturers are on the market. These cameras find application in many different disciplines and it is expected that they become a standard sensor device in image processing.

In the past decades a range of alternative depth estimation approaches have been proposed. The approaches typically operate with standard color or grayscale cameras without any active lighting. Image features are used to derive the distance to objects in the scene. Popular methods are Shape-from-Shading, Shade-from-Focus and Shape-from-Silhouette. Even more advanced methods learn common depth distributions in scenes and transfer this knowledge onto new scenes.

The use of color images is extremely widespread in computer vision, since color cameras are cheap, ubiquitous and provide valuable information. Thus, depth cameras are often used in conjunction with color cameras. Methods exist to calibrate and register a color and a depth camera and to map the measurements afterwards. This mapping is performed by relying on the measured distance to an object. Measurement noise and inaccuracies therefore affect the mapping and lead to errors. Additionally, the cameras have a different point of view and hence do not share the same view, which causes holes in the mapping for example at edges of objects. In order to overcome these problems, a monocular combination of a color and a depth imaging chip was developed at the ZESS. A beam splitter allows both imaging chips to share the same view onto the scene, which renders the mapping of both images unnecessary. This 2D/3D camera is named MultiCam and allows a depth independent registration of both modalities.

1.1 Limitations of 2D/3D Imaging and Contributions

Much progress has been made in increasing the lateral resolution, the sensitivity and the applicability of ToF-based depth imaging chips, but several limitations still persist. High reliability of depth imaging can only be achieved indoors under controlled lighting conditions. Most depth cameras cannot operate in highly illuminated surroundings especially in sunlight. Mechanisms were developed allowing depth cameras to work outdoors, but this reduces the measurement quality achievable.

A related limitation of depth cameras is their measurement range. Many indoor scenes contain only short distances of a few meters. However, higher distances occur in common scenes in professional environments or outdoors.

The distance limit of available depth cameras lies between 1 and 10 meters depending on the device. This reduces the applicability of depth cameras in many situations.

The resolution of the depth imaging chips is a widely discussed other limitation of depth cameras. The first evolution of depth imaging chips consisted of only 64×48 pixels, which did not allow a reconstruction of typical scene geometries. At the time of writing depth imaging chips with 160×120 or 200×200 pixels are available and chips with higher resolution are in the experimental stage. In general, higher resolutions lead to smaller pixels, which require a higher sensitivity to achieve the same measurement quality. This is the main limiting factor of the resolution of depth imaging chips.

The mentioned limitations are addressed in this thesis as follows: On the technical side two concepts for novel active lighting devices are presented to facilitate depth imaging for medium-range outdoor scenes up to 75 meters, depending on the opening angle, and for long-range scenes up to 150 meters. Processing steps for these imaging devices aiming at scene observation are introduced. In particular, different approaches on how to fuse high resolution color images with low resolution depth maps without introducing false information are discussed.

1.2 Thesis Outline

The operating principles of depth cameras available as commercial products as well as research prototypes are reviewed in Chapter 2. Their capabilities and limitations are compared by means of a number of evaluation setups. The focus lies in comparing Time-of-Flight cameras manufactured by the company PMD Technologies (PMDTec) and its competitor SoftKinetic as state-of-the-art depth imaging technology to the Microsoft Kinect, which is based on a rivaling structured light technology.

On the basis of this characterization of current depth cameras, the theory of a common branch of depth imaging chips named Photonic Mixer Device (PMD) is explained in detail in Chapter 3 and the behavior of PMD chips is analyzed. The ZESS MultiCam, a monocular 2D/3D camera, with which most experiments for this thesis were conducted, is introduced in conjunction with novel lighting devices to facilitate depth measurement for distances up to 150 meters. Additionally, an approach to gain absolute depth values for long-range depth measurements with PMD chips is demonstrated. Methods to improve the depth images in order to obtain accurate depth measurements are reviewed in Chapter 4 and a method to remove fixed pattern noise for PMD chips is introduced. These camera calibration methods require a sequence of measurements and a specific procedure to estimate camera parameters, which are sometimes not available or not possible to obtain. Therefore, a novel method to derive at least the most important camera parameters while requiring just a single depth image is presented.

One of the most common video processing tasks is background subtraction. The goal is to identify the foreground of a video and the mostly static background. In Chapter 5 the standard approach for background subtraction of color videos is adapted to 2D/3D videos, i.e. color videos accompanied with low resolution depth maps. The method is expressed in a probabilistic framework, which on the one hand allows an improved handling of dynamic color video effects like shadows as well as foreground background similarities and on the other hand ensures that invalid depth measurements and noise are treated appropriately.

However, effects of the significantly lower resolution of the depth maps remain an issue. Therefore, a set of recently proposed super-resolution methods for 2D/3D images is compared in Chapter 6. Since these methods are computationally expensive and may introduce inaccurate estimates in the depth maps, novel approaches to perform the super-resolution jointly with the main task instead of applying it as a pre-processing step are investigated. Firstly, a joint motion compensation and super-resolution method to reduce the effects of involuntary camera motion is proposed. Secondly, a framework to estimate high resolution depth maps iteratively while performing a 2D/3D segmentation is detailed and evaluated.

Finally, in Chapter 7 a global multi-camera and multi-modal tracking approach is introduced. The method is able to utilize any number and type of cameras. Information from color cameras is used by projecting the image onto a possible target modeled with an ellipsoid in 3D and comparing it to the expected appearance. In order to exploit depth information, the observed density at each line of sight is modeled and based on these models the density of a possible target is estimated. Several experiments were conducted to illustrate the capabilities of the tracking approach. The thesis closes with a conclusion in Chapter 8 and an overview of future research topics in the area of 2D/3D imaging.

2 Depth Camera Assessment

The driving question of this chapter is how competitive cheap consumer depth cameras, namely the Microsoft Kinect and the SoftKinetic DepthSense, are compared to state-of-the-art Time-of-Flight depth cameras. Several depth camera models from different manufactures are put to the test on a variety of tasks in order to judge their respective performance and to reveal their weaknesses. The evaluation will concentrate on the area of application for which all cameras are specified, i.e. near field indoor scenes. The characteristics and limitations of the different technologies as well as the available devices are discussed and evaluated with a set of experimental setups. In particular, the noise level and the axial and angular resolutions are compared. Additionally, refined formulae to generate depth values based on the raw measurements of the Kinect are presented.

2.1 Depth Camera Overview

Depth or range cameras have been developed for several years and are available to researchers as well as commercially for certain applications for about a decade. PMD Technologies (PMDTec), Mesa Imaging, 3DV Systems and Canesta were the companies driving the development of Time-of-Flight (ToF) depth cameras. In recent years additional competitors like Panasonic, SoftKinetic or Fotonic announced or released new models.

The cameras produced by all these manufacturers have in common that they illuminate the scene with infrared light and measure the time until the light is received. There are two main principles of operation: pulsed light and continuous wave amplitude modulation. The former is limited by having to measure very short time intervals in order to achieve a distance resolution which corresponds to a few centimeters in depth (e.g. ZCam of 3DV Systems). The continuous wave modulation approach avoids this by measuring the phase shift between emitted and received modulated light, which corresponds directly to the time of flight and in turn to the depth. However, ambiguities in form of multiples of the modulation wavelength may occur here.

In the past the ToF imaging sensors suffered from two major problems: a low resolution and a low sensitivity resulting in high noise levels. Additionally, background light caused problems when used outdoors. Currently, ToF imaging chips reaching resolutions of up to 200×200 pixels are on the market and chips with 352×288 pixels are in development. Moreover, for a few years some ToF chips have featured methods to suppress ambient light (e.g. Suppression of Background Illumination - SBI).

Other depth cameras or measuring devices, such as laser scanners or structured light approaches, were not able to provide (affordably) high frame rates for full images with a reasonable resolution. This was true until in 2010 Microsoft (PrimeSense) released the Kinect. Instead of relying on a pattern varying in time as widely applied previously, it works with a fixed irregular pattern consisting of a large number of dots produced by an infrared laser LED and a diffractive optical element. The Kinect determines the disparities between the emitted light beam and the observed position of the light dot with a two megapixel grayscale imaging chip. The identity of a dot is determined by utilizing the irregular pattern. It is assumed that the depth of a local group of dots is calculated simultaneously, but the actual method remains a secret up until today. Once the identity of a dot is known the distance to the reflecting object can be easily triangulated. In addition to the depth measurements, the Kinect includes a color imaging chip as well as microphones.

Given the low cost of the Kinect as a consumer product and the novelty as well as the non-disclosure of its functional principle, the reliability and accuracy of the camera should be evaluated. Instead of an investigation of a specific application, the approach taken to judge the performance of the Kinect is to develop a set of experimental setups and to compare the results of the Kinect to state-of-the-art ToF depth cameras.

The performance of ToF cameras using the Photonic Mixer Divece (PMD) was widely studied in the past. Noteworthy are for example [55, 11]. The measurement quality at different distances and using different exposure times is evaluated. Lottner et al. discuss the influence and the operation of unusual lighting geometries in [45], i.e. lighting devices not positioned symmetrically around the camera in close distance. In [66] depth cameras from several manufactures are compared, which are PMDTec, Mesa Imaging and Canesta. The application considered is 3D reconstruction for mobile robots. And in [2] PMD cameras are compared to a stereo setup. They use the task of scene reconstruction to judge the performance of both alternatives. The most closely related paper is [61], in which two ToF cameras are compared to the Kinect and to a laser scanner. The application in mind is navigation

Figure 2.1: The depth cameras involved in the comparison. In the top row
left to right are the MicroSoft Kinect, two ZESS MultiCams, the
PMDTec 3k-S displayed and in the bottom row the SoftKinetic
DepthSense 311 and finally the PMDTec CamCube 41k.

for mobile robots and the methodology is the reconstruction of a 3D scene
with known ground truth.

The following comparison was previously published in [L8] and involves
different commercially available depth cameras, some of which are shown in
Fig. 2.1 as well as several versions of our MultiCam. The Microsoft Kinect
and the SoftKinetic DepthSense as recent consumer depth cameras compete
with two established Time-of-Flight cameras based on the Photonic Mixer
Device (PMD) by PMDTec.

2.1.1 Microsoft Kinect

The Microsoft Kinect camera generates an irregular pattern of dots (actually,
a sub-pattern is repeated 3×3 times) with the help of a diffractive optical
element and an infrared laser diode. It incorporates a color and a two
megapixel grayscale chip with an IR filter, which is used to determine the
disparities between the emitted light dots and their observed position. In
order to triangulate the depth of an object in the scene, the identity of an
observed dot on the object must be determined. This can be performed with
much more certainty with the irregular pattern than with an regular pattern.
The camera is built with a 6 mm lens for the color chip and an astigmatic
lens for the grayscale chip, which skews the infrared dots to ellipsoids. These
deformations provide a depth estimate and together with the triangulation
a depth map is calculated. In the standard mode the depth map contains
640×480 pixels and each pixel is a raw 11-bit integer value. The depth

values describe the distance to the imaginary image plane and not to the focal point. There are currently two formulae to calculate the depth in meters publicly known, cf. [49]. An integer raw depth value d is mapped to a depth value in meters with a simple formula by

$$\delta_{simple}(d) = \frac{1}{-0.00307d + 3.33} \ . \tag{2.1}$$

A more precise method based on a higher order function is be given by

$$\delta_{tan}(d) = 0.1236 \cdot \tan\left(\frac{d}{2842.5} + 1.186\right) \ . \tag{2.2}$$

Since the depth map has about $300k$ pixels, calculating the latter formula 30 times per second can be challenging or even impossible, especially for embedded systems.

Using the translation unit described in Section 2.2.1 refined formulas have been determined:

$$\delta_{simple}^{refined}(d) = \frac{1}{-0.8965 \cdot d + 3.123} \tag{2.3}$$

$$\delta_{tan}^{refined}(d) = 0.181 \cdot \tan\left(0.161 \cdot d + 4.15\right) \ . \tag{2.4}$$

See Section 2.2.1 for a comparison of these formulae.

2.1.2 PMDTec CamCube 41k

The CamCube 41k by PMDTec, cf. [51], contains a 200×200 pixel PMD chip and includes two lighting units. Modulated infrared light with frequencies up to 21 MHz is emitted and the phase shift between the emitted and received light is calculated. The phase corresponds to the distance of the reflecting object and it is determined using the so-called four phase algorithm. For this algorithm four phase images P_1 to P_4 are recorded at different phase offsets and with the arc tangent relationship the phase difference can be retrieved as

$$\Delta\varphi = \text{atan2}\left(P_2 - P_4, P_1 - P_3\right) \ . \tag{2.5}$$

The distance can be derived from the phase difference with

$$\delta(\Delta\varphi) = \Delta\varphi \cdot \frac{c}{4\pi \cdot \nu} \ , \tag{2.6}$$

where c is the speed of light and ν is the modulation frequency. More details will be discussed in Section 3.1.

The CamCube features a method to suppress background illumination called SBI to allow for outdoor imaging. It provides the possibility to synchronize the acquisition of images with the means of a hardware trigger. A wide variety of different lenses can be mounted on the camera due to the standard CS-mount adapter. The camera is connected to a computer via USB.

2.1.3 PMDTec 3k-S

The 3k-S PMD camera is a development and research version from PMDTec and it employs an older PMD chip with only 64×48 pixels. It features a SBI system and contains a C-mount lens adapter and uses firewire (IEEE-1394) to communicate with the computer. This PMD camera is known to be significantly less sensitive than cameras with newer PMD chips even though the pixel pitch is $100\,\mu m$ compared to $45\,\mu m$ of the 41k PMD chip.

2.1.4 PMDTec 100k

PMDTec is developing depth imaging chips of higher resolution and a version of a 100k PMD chip with 352×288 pixels and a pixel pitch of $17.5\,\mu m$ is tested as well. The chip features also a SBI system and is electronically similar to earlier versions. However, it contains inhomogeneities and certain deficiencies which will be discussed later on.

2.1.5 Softkinetic DepthSense 311

The newly established company SoftKinetic released a depth imaging camera named DepthSense 311 in 2012, which includes an additional color camera in a binocular setup and microphones. The camera is also based on the ToF principle and modulates infrared light with frequencies between 14 and 16 MHz. The modulation frequency is continuously changed and 500 phase images are acquired per second resulting in up to 60 fps. The lateral resolution of the depth map is 160×120 pixels and color images of 640×480 pixels are delivered. The camera is connected via USB and contains a fixed lens with an opening angle of 57.3 degrees in width for the depth chip.

2.2 Experimental Evaluation

In this section the evaluation methods and the most notable results of the comparison will be discussed. In the first part in Section 2.2.1, the radial resolution in terms of precision and accuracy of all cameras will be compared. For the second set of experiments only the CamCube will serve as a reference for the Kinect, since only the different technologies ToF and triangulation are compared. In order to make the results comparable an appropriate lens has to be chosen for the CamCube. Since the Kinect uses a fixed 6 mm lens and the grayscale chip has a resolution of 1280×1024 (only 640×480 depth pixels are transmitted) with a pixel pitch of $5.2\,\mu m$, this results in a chip size of $6.66\,mm \times 5.33\,mm$. The CamCube has a resolution of 200×200 pixels with a pixel pitch of $45\,\mu m$ resulting in a chip size of $9\,mm \times 9\,mm$. Therefore, the corresponding lens for the CamCube would have a focal length of 8.11 mm for the width and about 10.13 mm for the height. As a compromise a lens with a focal length of 8.5 mm was chosen.

2.2.1 Depth Accuracy Evaluation

All cameras were mounted on a translation unit, which is able to position the cameras at distances between 50 cm and 5 meters from a wall with a positioning accuracy better than 1 mm. The cameras were leveled and were pointing orthogonally at the wall. 100 images were taken per position with a step size of 1 cm, which resulted in 45000 images per camera. The same lens, the same modulation frequency of 20 MHz as well as the same exposure times (5 ms for distances below 2.5 meters and 10 ms for higher distances) were used for all PMD based cameras. In Fig. 2.2 single depth measurements, the estimated standard deviation (SD) as well as the average distance error of the measurements to the ground truth after a linear correction are shown for a single pixel of each evaluated camera. The average error is computed by performing a linear regression for a subset of depth measurements (2 to 4 meters) and correction all depth measurements afterwards. This removes constant and linear errors in the measurements, which are caused by systematic errors or inaccuracies in the setup. Similar plots were made for different pixels to ensure that the results are representative for the cameras.

Here the CamCube shows measurement errors for small distances due to overexposure and both PMD based ToF cameras display the wobbling behavior as previously discussed, e.g. in [34]. The distance error for all cameras is comparable in the optimal region of application $(2 - 4m)$ with a slight advantage for the Kinect. More complex calibration methods exist for PMD based cameras, see [41] or [58], which are able to reduce the distance error further. The estimated standard deviation of the distance

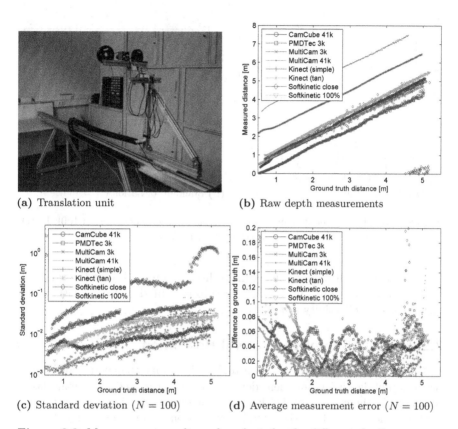

(a) Translation unit   **(b)** Raw depth measurements

(c) Standard deviation ($N = 100$)   **(d)** Average measurement error ($N = 100$)

Figure 2.2: Measurement results and analysis for the different depth cameras performed with the translation unit.

measurements based on 100 frames shows significant differences. The Kinect displays a low variance for short distances but higher noise levels than e.g. the CamCube for distances larger than two meters. The variance of the PMDTec 3k camera is higher due to its limited lighting system and its low sensitivity. The SoftKinetic DepthSense is in this respect slightly inferior to other state-of-the-art depth cameras. The experimental 100k PMD chip is evaluated in Fig. 2.3. The higher measurement noise is a consequence of the much smaller pixel size. The pixels are also very easily overexposed. Consequently, a small exposure time of 1 ms was applied. Nevertheless, an overexposure is observed for distances closer than 1.5 meters. Shorter

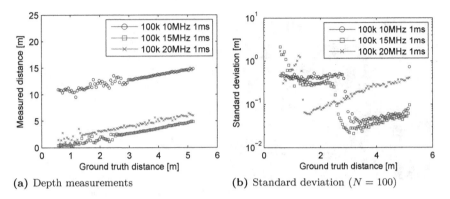

(a) Depth measurements   **(b)** Standard deviation ($N = 100$)

Figure 2.3: Measurement results and analysis for the experimental 100k PMD chip performed with a translation unit.

exposure times will reduce this minimal distance but will limit the range of the camera. Therefore, the 100k PMD chip is limited to scenes with small distances and reflectivity differences.

2.2.2 Estimation of the Lateral Resolution

A 3D Siemens star, see Fig. 2.4, is a tool to determine the angular or lateral resolution of depth measurement devices. In [8] it was used to compare laser scanners. In the context of depth cameras it promises insights, in particular for the Kinect, for which the effective resolution is not known. The lateral resolution r of an imaging device can be calculated as

$$r = \frac{\pi d M}{n} \tag{2.7}$$

with n being the number of fields of the star (here 12 and 24 respectively), d being the ratio of the diameter of an imaginary circle in the middle of the star containing incorrect measurements to the diameter M of the star.

For the 3D Siemens stars frames were taken at different distances and in Fig. 2.4 the respective parts of one set of the resulting images are shown. In theory, the CamCube has an opening angle of 55.8 degrees with a 8.5 mm lens, which leads to an angular resolution of 0.28 degrees. Using the 3D Siemens stars in one meter distance an estimate for the angular physical resolution of the CamCube is 0.51 cm, which corresponds to 0.29 degrees and confirms the theoretical value.

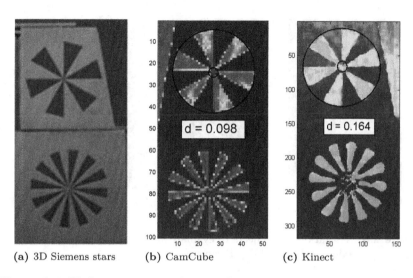

(a) 3D Siemens stars   (b) CamCube   (c) Kinect

Figure 2.4: 3D Siemens stars with 20 cm diameter and measurement results in 1 meter distance.

The Kinect has an opening angle of 58.1 degrees and with 640 pixels in width it has a theoretical angular resolution of 0.09 degrees (0.12° in height). In practice an angular resolution of 0.85 cm and 0.49 degrees was determined. This corresponds to a resolution of 118 pixels in width. The significant difference is due to the fact that multiple pixels are needed in order to generate one depth value (by locating the infrared dot). Even though the Kinect contains a two megapixel grayscale chip and transfers only a VGA depth map, this still does not compensate the need of multiple pixels in order to locate the dots. Additionally, the Kinect performs to our knowledge either a post-processing or utilizes regions of pixels in the triangulation, which may lead to errors at boundaries of objects.

This observation agrees with estimates that the dot pattern consists of about 220 × 170 dots which can be interpreted as the theoretical limit of the lateral resolution.

2.2.3 Depth Resolution Test Objects

Fig. 2.5 shows three objects to visualize the angular and axial resolution of the depth cameras. The first object consists of a ground plane and three

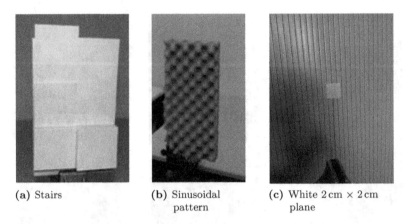

(a) Stairs   (b) Sinusoidal   (c) White 2 cm × 2 cm
                            pattern   plane

Figure 2.5: Resolution test objects to evaluate and visualize the angular and
axial resolution of depth cameras.

6 cm × 6 cm cuboids of different heights of 3, 9 and 1 mm. The second object
has a surface close to a sinusoidal formed plane with an amplitude of 1.75 cm
and a wave length of 3.5cm. Moreover, a 2 cm × 2 cm white plane mounted
on a 0.5 mm steel wire was placed in some distance to a wall. Then the
depth cameras were positioned at different distances to the plane and it was
checked whether they were able to distinguish between the plane and the
wall.

In Fig. 2.6 some results for the cuboids are shown, for which 10 depth maps
were averaged. Both cameras are able to measure the different distances
with high accuracy in one meter distance. At 1.5 meters distance the
precision decreases and at 2 meters both cameras cannot resolve the pattern
reliably. In Fig. 2.7 a rendering of the sinusoidal structure is given. Again
both cameras are able to capture the pattern correctly, but the detail of
preservation is higher for the Kinect.

The experiment with the small plane yields surprising results. For the
CamCube the 2 cm × 2 cm plane stays visible with correct depth value even
in 4.5 m distance. The plane has a size of only 0.7 pixels when the camera is
placed at this distance, but this is still enough to gain a correct measurement.
The pixel will observe a mixture of signals with different phases, but the one
coming from the small plane is the strongest and therefore the measurement
still yields sufficiently reliable values. The Kinect displays a completely
different behavior. Here a second camera with an optical IR filter was

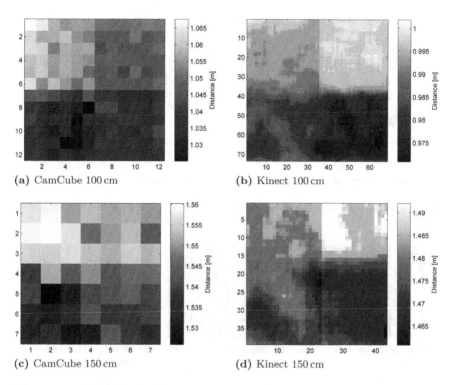

(a) CamCube 100 cm

(b) Kinect 100 cm

(c) CamCube 150 cm

(d) Kinect 150 cm

Figure 2.6: Cuboids of different heights recorded using the Kinect and the Camcube. 10 frames were averaged for each distance.

employed to observe the infrared dots on the plane. In 1.75 meters distance the small plane is invisible to the Kinect, as the number of dots on the plane is less than five. In 1.5 meter distance the plane is visible in about 50% of the cases depending on the lateral position of the plane, for an example see Fig. 2.8. In one meter distance the plane is visible and correctly measured all the time with about 10 − 20 dots on the plane. The explanation for this behavior is the same as for the 3D Siemens stars.

2.2.4 Angular Dependency of Measurements

Since the measurements of the Kinect are based on triangulation, it is doubtful that objects can be measured accurately at all angles. To evaluate

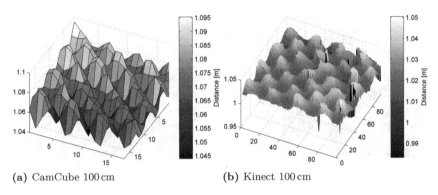

(a) CamCube 100 cm   (b) Kinect 100 cm

Figure 2.7: Sinusoidal structure measured with both cameras in 1 m distance.

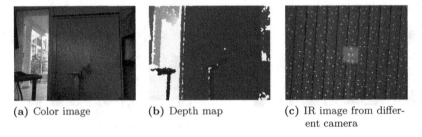

(a) Color image   (b) Depth map   (c) IR image from different camera

Figure 2.8: Measurement result of the Kinect for the small plane in 1.5 m distance.

the range of angles resulting in accurate measurements the camera is moved horizontally and a plane is installed in a fixed distance to the camera path. Angles from −40 to −110 degrees and from 40 to 110 degrees with a step size of 5 degrees are applied and the camera is positioned with offsets from −1 to 1 meter using a step size of 10 cm. High accuracy positioning and rotation devices are used for this purpose. This leads to a total number of 30 × 21 images. For each image the measurement quality is evaluated and grades are assigned: All pixels valid, more than 80% valid, more than 20% valid and less than 20% valid.

In Fig. 2.9 the results for the test setup to identify difficulties in measuring sharp angles are shown. Measuring errors for angles up to 20 degrees less than the theoretical limit, i.e. the angle in which the front side of the plane

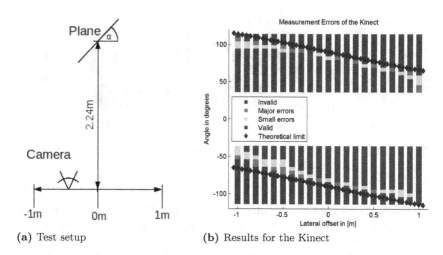

(a) Test setup   **(b)** Results for the Kinect

Figure 2.9: Setup to test the ability of the Kinect to measure planar objects with different angles and positions relative to a camera path and results.

is invisible, are encountered. It is noteworthy that the left side of the depth map is affected significant higher than the right side. This is where the grayscale camera is located and therefore, the angle under which the incident light strikes the plane is here smaller than on the right side.

2.2.5 Limitations

In this evaluation and in previous experiments the different types of cameras displayed different weaknesses. The Kinect showed problems with dull (LCD monitors) or shiny surfaces or surfaces under a sharp viewing angle. Obviously, mounting the Kinect is relatively difficult and the lens is not exchangeable, which limits its application. Different lenses in combination with different diffractive optical elements might for example allow for larger distances. These drawbacks might be solved in different hardware implementations, but the largest problems are caused by systematic limitations. A significant part of a typical depth map contains no measurements due to shading: certain regions of the objects seen by the grayscale camera are not illuminated by the IR light beam. Depending on the application these areas can cause huge problems. In Fig. 2.10 a challenging test scene is shown. Here black indicates invalid measurements in the depth map for the Kinect.

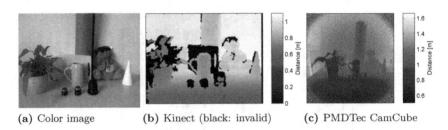

(a) Color image   (b) Kinect (black: invalid)   (c) PMDTec CamCube

Figure 2.10: Resulting depth maps of a difficult test scene using the Kinect and the PMDTec CamCube.

Daylight is another source of problems. Since the grayscale chip of the Kinect uses an optical filter only infrared light disturbs the measurements. Therefore, a high power infrared LED with a peak wavelength at 850 nm and an infrared diode with corresponding characteristics have been tested to give an impression at which levels of infrared ambient light the Kinect can be used. It has been determined that measuring errors occur for an irradiance of $6 - 7\,\mathrm{W/m^2}$ depending on the distance. For comparison: sunlight at sea level has an irradiance of about $75\,\mathrm{W/m^2}$ for wavelengths between 800 and 900 nm.

The limitations of PMD based ToF cameras are mainly motion artifacts, which occur when objects move significantly during the acquisition of the four phase images. Another problem are mixed phases, which are produced when a pixel observes modulated light with different phase shifts due to reflections or borders of objects inside a pixel. Additionally, the low resolution and the higher power requirements limit the application of ToF cameras to some degree.

2.3 Summary

In this chapter a consumer depth camera, the Microsoft Kinect working with a novel depth imaging technique, is compared to state-of-the-art continuous wave amplitude modulation Time-of-Flight cameras. A set of experimental setups was devised to evaluate the respective strengths and weaknesses of the cameras as well as the underlying technology.

It was found that the new technique as well as the available device poses a strong competition in the area of indoor depth imaging with small distances. Only the problems caused by the triangulation, namely shading due to different viewpoints, measuring difficulties of sharp angles and measuring of small structures are major weaknesses of the Kinect.

The Kinect as well as the DepthSense are not able to measure distances under high illumination, especially in sunlight. The PMD Time-of-Flight imaging chips with SBI are able to measure distances in sunlight, but with significantly lower exposure times, i.e. one millisecond, leading to a reduced quality. Therefore, acquiring full frame depth measurements at high frame rates in an outdoor environment or for longer distances is the domain of ToF chips with SBI mechanism up until today. For indoor scenes higher resolutions like the currently developed 100k PMD chip by PMDTec may level the playing field again.

In the following chapters PMD chips will be used as a basis to develop depth imaging techniques for medium and long distances due to the severe limitations of the other devices.

3 PMD Imaging

In this chapter new approaches and discoveries associated with the Photonic Mixer Device (PMD) technology are discussed. The chapter begins with an overview of the PMD operation principles and the theory including the Suppression of Background Illumination (SBI) mechanism in Section 3.1. In general the depth measurement accuracy and resolution depend on the modulation frequency. The effects of the modulation frequency as well as sources of measurement errors, most notably the intensity related error and a newly discovered SBI induced error, are discussed in this section and were previously published in [L9].

Afterwards a new version of the ZESS MultiCam, a monocular 2D/3D camera, is introduced in Section 3.2. Many experiments presented in this thesis were conducted with this device. Novel active lighting systems to be used in conjunction with this camera are introduced in Section 3.3. Firstly, a medium-range lighting for distances up to 70 meters (depending on the opening angle) as well as a long-range illumination device for distances above 100 meters are presented. In later chapters methods to facilitate the application of these lighting devices are introduced. Approaches for phase-unwrapping are investigated in Section 3.4, which are necessary for long-range measurements since PMD is based on continuous wave amplitude modulation and the distance measurements are ambiguous. If only distances lower than the unambiguous range are present in the observed scene, such methods are obviously not necessary. However, for larger distances the precision of high modulation frequencies and the unambiguity range of low frequencies are desired and hence phase unwrapping is necessary to retrieve absolute distances.

3.1 PMD Operation Principles

A PMD chip measures distances based on the Time-of-Flight (ToF) principle. A scene is illuminated and the time until the emitted light is received is measured in each pixel. In general, this can be achieved with pulsed light and measuring the time between casting the light and receiving it directly.

However, very short times in the region of tens or hundreds of pico seconds have to be measured accurately depending on the desired distance accuracy. In order to reduce the complexity of imaging chips, the time of flight can be measured indirectly as well. PMD technology operates by measuring the phase difference between emitted amplitude modulated near infrared light and the received light. Usually four samples with different phase offsets are acquired and the phase shift is calculated with the four phase algorithm. In the following the theory is described in more detail.

The modulated infrared light emitted at time t is received with phase shift $\phi = \phi_d + \phi_{off}$ containing a part ϕ_d depending on the ToF and an artificial phase shift ϕ_{off}. The received light is modeled with

$$L(t, \phi) = L_0 + \hat{L} \cdot \sin(\omega t + \phi) \tag{3.1}$$

with an amplitude $\hat{L}$, an intensity offset $L_0 \geq \hat{L}/2$ and a modulation frequency $\nu = \omega/2\pi$. A PMD pixel consists of two channels denoted with A and B. A channel $\Gamma \in \{A, B\}$ performs an integration over the received light signal

$$U_\Gamma(\phi) = \int_{t=0}^{T} g_\Gamma(t) \cdot L(t, \phi) dt \tag{3.2}$$

with a correlation function $g_\Gamma(\cdot)$, which accounts for the observed amount of light and the imaging parameters. This correlation function is often assumed to be sinusoidal too, but it was shown in [47] that its shape is much closer to a rectangular signal. Therefore, the function is modeled here as

$$g_\Gamma(t) = \begin{cases} \lambda_\Gamma & , \quad \text{if } t \cdot \nu \mod 1 \leq 0.5 \\ \kappa_\Gamma & , \quad \text{else} \end{cases} . \tag{3.3}$$

Here λ_Γ and κ_Γ with $\lambda_\Gamma \neq \kappa_\Gamma$ are hardware dependent parameters. Many PMD derivations assume that $\lambda_A = \kappa_B = 1$ and $\lambda_B = \kappa_A = 0$ but this is neither a necessary nor a valid assumption. Now the integration can be performed as

$$U_\Gamma(\phi) = \int_{t=0}^{T} g_\Gamma(t) \cdot L(t, \phi) dt = \left[tL_0 g_\Gamma(t) - \frac{\hat{L}}{\omega} g_\Gamma(t) \cdot \cos(\omega t + \phi) \right]_0^T \tag{3.4}$$

and with the approximation $T = n/\nu$ for a $n \in \mathbb{N}$ this can be written as

$$U_\Gamma(\phi) \approx n \left[tL_0\lambda_\Gamma - \frac{\hat{L}}{\omega}\lambda_\Gamma \cdot \cos(\omega t + \phi) \right]_0^{1/2\nu}$$

$$+ n \left[tL_0\kappa_\Gamma - \frac{\hat{L}}{\omega}\kappa_\Gamma \cdot \cos(\omega t + \phi) \right]_{1/2\nu}^{1/\nu} . \quad (3.5)$$

Now the terms can be rewritten as

$$U_\Gamma(\phi) \approx n \left(\frac{L_0}{2\nu}(\lambda_\Gamma + \kappa_\Gamma) - \frac{\hat{L}}{\omega}\lambda_\Gamma \cdot \cos\left(\frac{\omega}{2\nu} + \phi\right) \right.$$

$$\left. + \frac{\hat{L}}{\omega}\lambda_\Gamma \cdot \cos(\phi) - \frac{\hat{L}}{\omega}\kappa_\Gamma \cdot \cos(\phi) + \frac{\hat{L}}{\omega}\kappa_\Gamma \cdot \cos\left(\frac{\omega}{2\nu} + \phi\right) \right) \quad (3.6)$$

and simplified to

$$U_\Gamma(\phi) \approx n \left(\frac{L_0}{2\nu}(\lambda_\Gamma + \kappa_\Gamma) + \frac{2\hat{L}}{\omega}(\lambda_\Gamma - \kappa_\Gamma) \cdot \cos(\phi) \right) . \quad (3.7)$$

This integration result is digitized with gain $\hat{N}_\Gamma$ and offset N_Γ^0 at each channel Γ of a pixel leading to

$$N_\Gamma(\phi) = N_\Gamma^0 + n \cdot \hat{N}_\Gamma \left(\frac{L_0}{2\nu}(\lambda_\Gamma + \kappa_\Gamma) + \frac{2\hat{L}}{\omega}(\lambda_\Gamma - \kappa_\Gamma) \cdot \cos(\phi) \right) . \quad (3.8)$$

Now it can be easily shown that the following holds for phase shifts $\phi_{off} = 0, \pi/2, \pi, 3\pi/2$:

$$\frac{N_A\left(\phi_d + \frac{\pi}{2}\right) - N_B\left(\phi_d + \frac{\pi}{2}\right) - N_A\left(\phi_d + \frac{3\pi}{2}\right) + N_B\left(\phi_d + \frac{3\pi}{2}\right)}{N_A(\phi_d) - N_B(\phi_d) - N_A(\phi_d + \pi) + N_B(\phi_d + \pi)} = \tan(\phi_d) .$$
$$(3.9)$$

Once the distance related phase shift ϕ_d is recovered with the 8 samples, the distance d to the reflective object can be calculated as

$$d = \frac{\phi_d c}{4\pi\nu} \quad (3.10)$$

with c being the speed of light. In practice, there are numerous effects, which influence the phase estimation. Firstly, there is a delay between the generation of the modulation signal and the emittance of the light, which results in a constant phase offset. Secondly, the emitted light is not sinusoidal, since light cannot be shaped that precisely with the state-of-the-art LEDs for high modulation frequencies. This creates the phase dependent wobbling, described in e.g. [41]. Another phase estimation error is caused by the non-linear behavior of a PMD channel and it is referred to as intensity related error. All these sources of phase estimation errors are discussed in the subsequent sections.

The amplitude $\hat{L}$ of the modulated light can be retrieved based on the 8 samples when assuming $\hat{N}_A = \hat{N}_B = \hat{N}$ and $\lambda_A + \kappa_A = \lambda_B + \kappa_B$ with

$$\frac{1}{2}\sqrt{\left(N_A(\phi_d) - N_B(\phi_d) - N_A(\phi_d + \pi) + N_B(\phi_d + \pi)\right)^2}$$

$$\overline{+ \left(N_A(\phi_d + \pi/2) - N_B(\phi_d + \pi/2) - N_A(\phi_d + 3\pi/2) + N_B(\phi_d + 3\pi/2)\right)^2}$$

$$= n\hat{N}\frac{2\hat{L}}{\omega}(\lambda_\Gamma - \kappa_\Gamma)$$

$$= \hat{M} . \tag{3.11}$$

In the literature as well as in the following $\hat{M}$ will be referred to as modulation amplitude. Similarly, the intensity M of the infrared light observed by the PMD chip (grayscale value) is defined the average of all samples with

$$M = \frac{1}{4}\left(N_A(\phi_d + \pi/2) + N_B(\phi_d + \pi/2) + N_A(\phi_d + 3\pi/2) + N_B(\phi_d + 3\pi/2)\right.$$

$$\left. + N_A(\phi_d) + N_B(\phi_d) + N_A(\phi_d + \pi) + N_B(\phi_d + \pi)\right) . \tag{3.12}$$

3.1.1 Increasing the Modulation Frequency

In theory increasing the modulation frequency enhances the precision of the depth measurements, since the wavelength corresponds to smaller distances for larger modulation frequencies. However, producing modulated light with high power LEDs and high frequencies is difficult and leads to a deteriorated signal. Finding the optimal frequency for a given application is far from trivial and is also limited by the required frame rate and the available amount of active lighting. In the area of machine vision short distances and high precision are often required. In the following the behavior of PMD chips is investigated, when applying higher frequencies than the usual 20 MHz. All possible distances within the unambiguity interval should be evaluated. In order to maintain equal lighting conditions (ambient light, viewing angle, active lighting, ...) the evaluation is performed with a delay generator by

Figure 3.1: The depth camera evaluation board used in the experiments.

applying 25 different phase offsets. In Fig. 3.1 the test setup is shown and Fig. 3.2 depicts the estimated standard deviation (SD) at different phase offsets for a single pixel and for an array of pixels are displayed, when measuring with different modulation frequencies. Moreover, the estimated SD averaged over all 25 phase offsets is displayed for the whole test setup while applying different frequencies. The integration time is chosen so that a similar modulation amplitude is received for all modulation frequencies.

The results display a behavior as was expected with an optimal modulation frequency of 45 MHz. Higher frequencies do not reduce the measurement noise expressed in relative distances. Additionally, a shrinking of the optimal measurement area is observed for higher frequencies.

3.1.2 Effects of the Lighting Geometry

In this section the influence of the position of the lighting devices on the measurements is investigated. The discussion will give only a rough estimate of the effects and not an exhaustive treatment of the issue. In Fig. 3.3a the same setup as described in the previous section forms the basis of the experiments, but instead of changing the modulation frequency the lighting positions were modified while maintaining the position of the spots on the wall. The exposure times were altered to account for changed amounts of active light. The distance between the two lighting devices was changed and it was expected that the position close to the camera (12.5 cm) is superior to

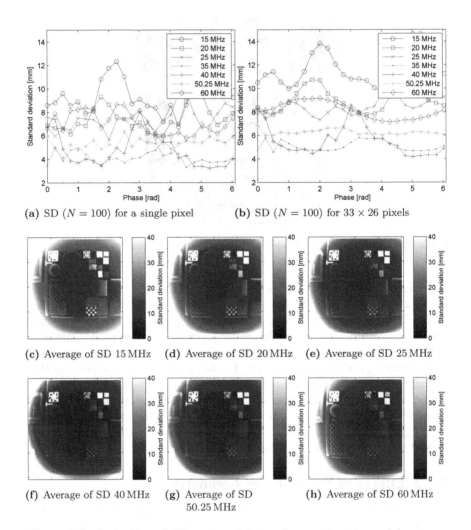

(a) SD ($N = 100$) for a single pixel   (b) SD ($N = 100$) for 33×26 pixels

(c) Average of SD 15 MHz   (d) Average of SD 20 MHz   (e) Average of SD 25 MHz

(f) Average of SD 40 MHz   (g) Average of SD   (h) Average of SD 60 MHz
                             50.25 MHz

Figure 3.2: Evaluation of different modulation frequencies using a delay generator to simulate 25 different distances while maintaining the exact
same lighting conditions.

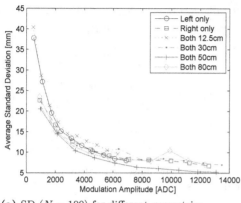

(a) SD ($N = 100$) for different geometries

(b) Custom camera setup

Figure 3.3: The effect of the position of the lighting system (two high power LEDs) is evaluated by comparing the standard deviation at different modulation amplitudes for a set of lighting positions.

larger distances. Additionally a single lighting device should also be better than two devices, since the light signal won't be affected by interference between the two light sources caused by the modulation. In contrast to the expectations, a distance of 50 cm between the lighting devices showed the best results. Moreover, only small differences were observed between all tested setups. Consequently, this allows for a relatively unrestricted placement of lighting devices enabling custom setups. Telecentric lenses and lenses with large focal lengths to capture small details at some distance to the camera are application examples. The lighting devices can then be positioned closely to the scene in order to focus the active light effectively, see Fig. 3.3b.

3.1.3 Signal Shape Related Error

One major cause of systematic measurement errors of PMD chips will be examined in this section. The PMD theory described in Section 3.1 assumes sinusoidal modulation signals, but in fact the lighting devices receive a nearly rectangular modulation signal and emit light with a signal shape, which is close to a rectangular signal for small frequencies and similar to a triangular signal for higher frequencies. This deformation of the rectangular modulation signal is caused by rise and fall times of the power transistor

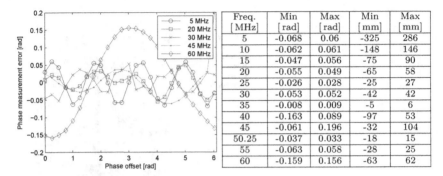

Freq. [MHz]	Min [rad]	Max [rad]	Min [mm]	Max [mm]
5	-0.068	0.06	-325	286
10	-0.062	0.061	-148	146
15	-0.047	0.056	-75	90
20	-0.055	0.049	-65	58
25	-0.026	0.028	-25	27
30	-0.053	0.052	-42	42
35	-0.008	0.009	-5	6
40	-0.163	0.089	-97	53
45	-0.061	0.196	-32	104
50.25	-0.037	0.033	-18	15
55	-0.063	0.058	-28	25
60	-0.159	0.156	-63	62

Figure 3.4: Plot and bounds for the nonlinear phase estimation error for selected modulation frequencies.

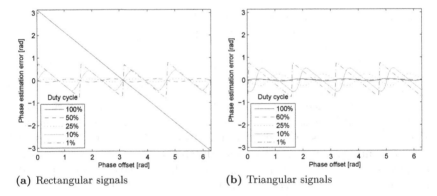

(a) Rectangular signals   (b) Triangular signals

Figure 3.5: Results of the nonlinear phase estimation error for simulated light signals of rectangular and triangular shape and for different duty cycles.

and the LED. The difference to a sinusoidal signal causes the wobbling error described in several publications. In [50] effects of the camera temperature are discussed, which can be caused by a changed dynamic behavior of the LEDs for increasing temperatures thus resulting in a changed signal shape.

In Fig. 3.4 results measuring the phase error for different modulation frequencies using a delay generator are shown. The images were acquired while ensuring that a similar modulation amplitude was observed. The difference between the 25 applied phase offsets and the measured signal phase averaged over 100 images after correction for a constant phase offset are plotted. Additionally, characteristics of the typical phase estimation error are given. These are specific to the lighting devices, but it is assumed that the results generalize well. In order to illustrate the cause to the error further, results for different simulated modulation signals and duty cycles are given in Fig. 3.5. Here rectangular and triangular signals were simulated and the phase estimation error using the four phase algorithm is calculated and a comparable behavior was observed.

The practical as well as the theoretical results show that the signal shape related error is usually limited to about one centimeter. Moreover, the difference of measurement errors between similar distances is very small and this leads to the conclusion that this error is not highly relevant for a wide range of applications, which have a rather small working area, e.g. possible measurement values between 1 and 2 meters. If deemed necessary, this error should be compensated for by applying a sinusoidal error model, see e.g. [46]. Previous publications overestimate the magnitude of this error because they do not distinguish clearly between the different error sources.

3.1.4 SBI Related Error

In this section the Suppression of Background Illumination mechanism (SBI) and a newly discovered measurement error caused by it are detailed. The SBI allows to some degree to utilize PMD chip outdoors by attempting to remove the L_0 part (called the DC part) of the received light. Thus, more active light can be incorporated in the integration and this decreases the stochastic error, which is caused by the imaging process as well as by modulation frequency jitter. In Fig. 3.6 two sets of measurements are visualized. In both sets raw PMD measurements of a wall in close distance (2 meters) were performed, while applying different exposure times. 100 images were acquired for each exposure time setting. In the top row the wall was lit with several halogen lamps and in the bottom row the level of ambient light was low. In each row the left plot shows the mean and the variance of the depth measurements. The right plot shows the raw PMD pixel values. A distance measurement using a PMD chip is conducted by acquiring four images with different phase shifts (0 deg, 90 deg, 180 deg, 270 deg). Since each PMD pixel consists of two channels this results in eight values. The third plot displays all acquired depth measurements.

Based on these measurements the effect of the SBI can be described. The SBI is triggered if the amount of received light on one channel exceeds a specific threshold (here approximately 36000 ADC). When the SBI is active, it charges both channels with the same charge. It can then be observed that

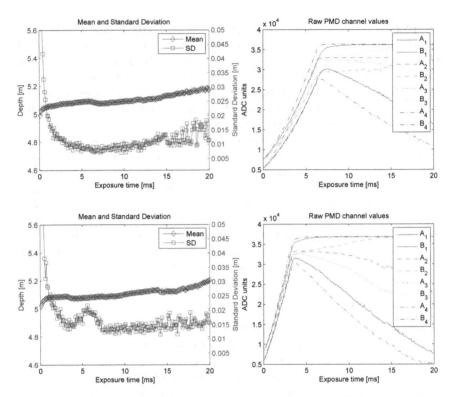

Figure 3.6: PMD measurements of a short distance using different exposure
times and under bright and dark ambient lighting conditions. 100
images were taken per exposure time step ranging from 0.1 ms to
20 ms. The top row shows the response under ambient light and
the bottom row without ambient light.

the charge of one channel remains in the threshold region in each acquisition
while the other channel is charged with the difference of the discharges of
both channels leading to lower ADC values. If a channel is fully charged
(at aprox. 5000 ADC), the PMD pixel enters the saturation region, which
affects the phase measurements severely.

In order to model a PMD channel with SBI both cases, i.e. channel A or
channel B reaching the threshold region first, are formulated separately. In

the following channel A shall be the channel observing the higher amount of light. Then channel A of a PMD channel can be modeled as

$$\tilde{N}_A(\phi) = N_A^0 + \hat{N}_A \left(n\frac{L_0}{2\nu}(\lambda_A + \kappa_A) + n\frac{2\hat{L}}{\omega}(\lambda_A - \kappa_A) \cdot \cos(\phi) \right. \tag{3.13}$$

$$\left. -\chi_{[0,\infty]}(n - \tilde{n}) \cdot (n - \tilde{n}) \left(\frac{L_0}{2\nu}(\lambda_A + \kappa_A) + \frac{2\hat{L}}{\omega}(\lambda_A - \kappa_A) \cdot \cos(\phi) \right) \right) .$$

Here $\chi_{[0,\infty]}(\cdot)$ denotes the indicator function with $\tilde{n} = \tilde{T}\nu$ describing the (discreet) time the when the SBI becomes active. The newly introduced term holds the channel in the threshold area, i.e. $\tilde{N}_A(\phi)$ reduces for $n \geq \tilde{n}$ to

$$\tilde{N}_A(\phi) = N_A^0 + \hat{N}_A \left(\tilde{n}\frac{L_0}{2\nu}(\lambda_A + \kappa_A) + \tilde{n}\frac{2\hat{L}}{\omega}(\lambda_A - \kappa_A) \cdot \cos(\phi) \right) . \tag{3.14}$$

The channel B modeled is this case as

$$\tilde{N}_B(\phi) = N_B^0 + \hat{N}_B \left(n\frac{L_0}{2\nu}(\lambda_B + \kappa_B) + n\frac{2\hat{L}}{\omega}(\lambda_B - \kappa_B) \cdot \cos(\phi) \right.$$

$$\left. -\frac{2\hat{L}}{\omega}(n - \tilde{n})\chi_{[0,\infty]}(n - \tilde{n}) \cdot (\lambda_A - \kappa_A)\cos(\phi) \right) . \tag{3.15}$$

When assuming that $\tilde{n}$ is equal for all four acquisitions, i.e. independent of the phase shift offsets $\phi_{off} = 0, \pi/2, \pi, 3\pi/2$, then the following holds:

$$\frac{\tilde{N}_A(\phi_d + \pi/2) - \tilde{N}_B(\phi_d + \pi/2) - \tilde{N}_A(\phi_d + 3\pi/2) + \tilde{N}_B(\phi_d + 3\pi/2)}{\tilde{N}_A(\phi_d) - \tilde{N}_B(\phi_d) - \tilde{N}_A(\phi_d + \pi) + \tilde{N}_B(\phi_d + \pi)}$$

$$= \frac{N_A(\phi_d + \pi/2) - N_B(\phi_d + \pi/2) - N_A(\phi_d + 3\pi/2) + N_B(\phi_d + 3\pi/2)}{N_A(\phi_d) - N_B(\phi_d) - N_A(\phi_d + \pi) + N_B(\phi_d + \pi)}$$

$$= \tan(\phi_d) . \tag{3.16}$$

However, this assumption is not valid, since the exact value of $\tilde{n}$ depends on the amount of incident light observed by each channel, which of course depends on the phase shift offsets ϕ_{off}. This leads to an imbalance comparable to a mixer imbalance in coherent demodulation. In the experiment with ambient light the SBI is triggered first for exposure times of about $3\,\text{ms}$ and at this time the measurement variance increases significantly.

Moreover, the following further characteristics can be observed:

- In theory the SBI eliminates all ambient light and the maximum possible exposure time is only affected by the amount of active light observed, or more precisely by the difference in active light both channels observe. However, in practice different efficiencies of both channels limit the integration time largely.

- It can be easily determined in most cases if the SBI has been triggered even for a single phase image, by observing the ADC values. This property can be exploited to set the exposure time automatically. A simple method is to simply count the number of pixels with active SBI and reduce the exposure time if this are more than 10% of all pixels.

3.1.5 Intensity Related Error

Another systematic error of PMD based depth cameras is independent of the phase offset, but exhibits a dependency on the intensity of the received light. This kind of error is observed most easily when measuring a checkerboard and was described in numerous publications, e.g. [34, 41]. In order to characterize this type or error, a dataset was acquired consisting of 100 images for integration times between 0.1 ms and 5 ms with a step size of

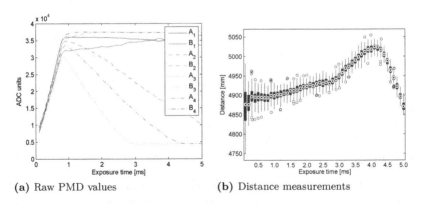

(a) Raw PMD values   (b) Distance measurements

Figure 3.7: Response curves of a PMD chip and depth measurements for different exposure times. The raw values for both channels and four phase offsets are shown on the left as well as a box plot of the resulting distance values on the right.

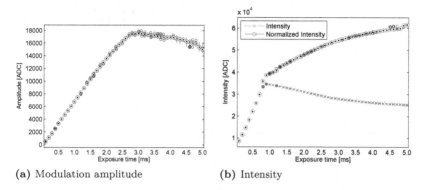

(a) Modulation amplitude   (b) Intensity

Figure 3.8: Resulting modulation amplitude and intensity values for different exposure times.

0.1 ms and under different indoor illumination settings. The raw values for both channels A and B as well as for the four phase images are displayed for one acquisition in Fig. 3.7a. The linear region of the PMD chip up to an integration time of 0.8 ms can be observed clearly. The SBI, which was detailed in the previous section, is active for longer integration times. The PMD chip enters the saturation region for one phase image when applying larger integration times than 2.5 ms. A box plot of the resulting depth measurements is displayed in Fig. 3.7b, where a 4 cm drift of the depth values can be observed, the absolute distance value of not of interest now. In Fig. 3.8 the corresponding modulation amplitude and intensity values are shown. In order to obtain increasing intensity values a normalization was applied as follows. If the value of one channel $\Gamma \in [A, B]$ of a PMD pixel in a phase image exceeds a threshold, e.g. $\gamma_{thres} = 36500$, a normalization term of $3 \cdot |A - B|$ is added to one channel. If the intensity response of a PMD chip is known for a given lens, a linear normalized intensity can be achieved.

During the experiments it was discovered that the intensity related error does not only depend on the modulation amplitude, but also on the level of ambient light. The responses of a PMD chip for different exposure times, phase offsets (using a delay generator) and different levels of ambient light were recorded. In Fig. 3.9 the resulting depth values were plotted for an observed modulation amplitude. The depth values for different phase offsets were previously normalized (displayed with different markers in the plot), so

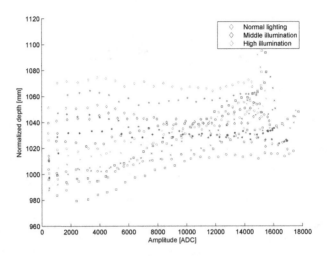

Figure 3.9: Depth measurements for different modulation amplitudes, phase offsets and levels of ambient light. Different markers denote different phase offsets. The resulting depth values were normalized, so that ideally a horizontal line should be observed.

that equal depth measurements should be obtained, and 100 measurements were averaged. The experiment was performed for different pixels and very similar results were obtained. The plot clearly shows a dependency of the depth measurements on the modulation amplitude, the phase and the level of ambient light.

3.2 ZESS MultiCam

At the ZESS a camera combining a color imaging chip with a PMD chip in a monocular setup (called MultiCam [52]) was developed. The most recent version is equipped with a Gigabit Ethernet interface, a three megapixel color CMOS chip and the 19k PMD chip from PMDTec. Both chips share the same lens utilizing a Bauernfeind prism with an integrated beam splitter. The heart of the camera is a Xilinx FPGA chip and the camera features a C-mount lens adapter. See Fig. 3.10 for pictures and specifications of the MultiCam. The lighting devices are modular and consist of chip LEDs with a maximum optical power of 3.5 Watts, which are available from Osram (SFH-4750). These LEDs have an emission peak at 860 nm and the active area is relatively dense in contrast to larger LED arrays. They also feature

MultiCam characteristics	
Interface	Gigabit Ethernet
Lens adapter	C-mount
Frame rate	12 fps (up to 80 fps with reduced 2D resolution)
Color chip	Aptina MT9T031
- Resolution	2048 × 1536
- Chip size	6.55 mm × 4.92 mm
Depth chip	PMDTec 19k (3k, 41k, 100k)
- Resolution	160 × 120
- Chip size	7.2 mm × 5.4 mm

Figure 3.10: The MultiCam, a 2D/3D monocular camera and its specifications.

low and symmetric rise and fall times of only 10 ns. The LEDs have a half-angle of 120 degrees but collimators with half-angles between 40 and 11 degrees can be applied and adjusted to cover the visible area of the MultiCam with a specific lens.

The frame rate of the camera is mainly limited by the color chip, which is able to provide 12 images of three megapixels size per second. However, it is possible to reduce the resolution of the color chip dynamically, leading to 93 fps at VGA resolution. For the most common applications it is useful to acquire depth maps synchronously to the color images, but for some applications high frame rates are more important. To that end the depth images can be transmitted independently of the color images. Additionally, a depth map can be calculated easily for each phase image received. With an exposure time of 1 ms frame rates of up to 400 fps are possible. Nevertheless, it should be noted that the acquisition time consisting of four phase images is at least about 10 ms, which must be taken into consideration for dynamic scenes. Motion compensation for PMD based cameras is discussed in [29].

3.3 Novel Lighting Devices

PMD imaging is restricted to small scenes with of the shelf hardware, covering up to 10 meters under favorable conditions. In this section novel lighting devices for non-standard ToF applications are presented. Firstly, a powerful lighting setup for wide opening angles and distances of up to 70 meters is introduced. Afterwards a long-range lighting setup for up to 150 meters and small opening angles relying on nearly parallel rays is discussed. The lighting devices were previously presented in [L10].

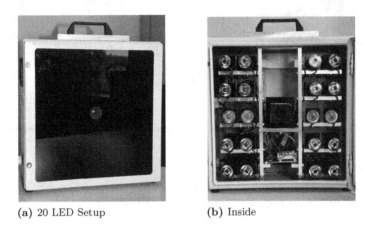

(a) 20 LED Setup   (b) Inside

Figure 3.11: Medium-range test setup with 20 LEDs.

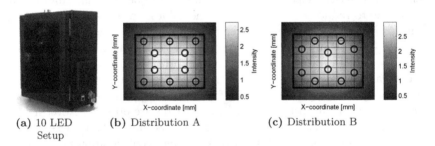

(a) 10 LED   (b) Distribution A   (c) Distribution B
   Setup

Figure 3.12: Small medium-range 2D/3D imaging device with 10 LEDs and simulations of possible intensity distributions depending on the positioning of the LEDs for the 19k PMD chip and a 12 mm lens in one meter distance.

3.3.1 Medium-Range Lighting

When applying PMD technology to outdoor scenes, the limitation in power of standard lighting devices is easily reached. Therefore, high-power adjustable lighting devices were constructed. Two prototypes, one with 10 and one with 20 LEDs each 3.5 Watts of continuous optical power each, are shown in Fig. 3.11 and Fig. 3.12. A collimator with a half angle of ±5.5 degrees is mounted on each LED (Osram SFH-4750) and the LEDs are adjusted to cover the observed area. Possible intensity distributions are simulated in Fig. 3.12 for the 10 LED device.

(a) Color image   (b) Depth (8 LEDs)

(c) Denoised depth   (d) Standard deviation ($N = 100$)

Figure 3.13: Medium-range outdoor test using only 8 LEDs. A modulation frequency of 20 MHz and an exposure time of 6 ms were used and the distance to the building is 50 meters.

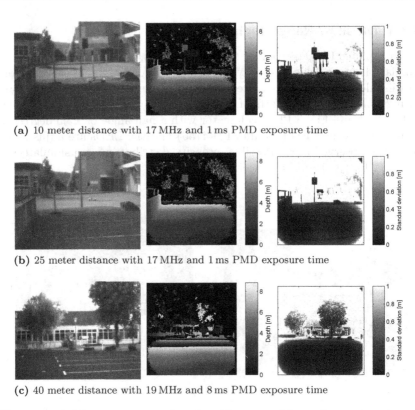

(a) 10 meter distance with 17 MHz and 1 ms PMD exposure time

(b) 25 meter distance with 17 MHz and 1 ms PMD exposure time

(c) 40 meter distance with 19 MHz and 8 ms PMD exposure time

Figure 3.14: Measurement results of a wide scene with a reference target containing poles in different distances on a sunny and a cloudy day acquired with the 20 LED setup and a 8.5 mm lens. Depicted are in each row from left to right the color image, a denoised depth map and the standard deviation based on 100 depth maps.

With these lighting systems scenes of up to 70 meters can be acquired, but the range obviously depends on the opening angle. Measurements of such a scene are given in Fig. 3.13 using only 8 LEDs and the distance to the building is about 50 meters. The results demonstrate the capabilities of this approach, but also show some weaknesses for small or highly structured objects due to the limited lateral resolution of the PMD chip. In order to investigate the medium-range lighting setup with 20 LEDs more thoroughly, experiments with a reference target containing poles of different diameters and reflectivity were conducted. Fig. 3.14 shows some results, taken under different lighting conditions and with a 8.5 mm lens leading to an opening angle of nearly 56 degrees for the PMD chip. A threshold based on the modulation amplitude (50 ADC) was applied to the depth maps in order to remove noise (dark blue). On a cloudy day higher exposure times can be applied leading to lower noise levels so that measurements of the reference target are possible even in 40 meters distance.

3.3.2 Long-Range Lighting

A special kind of illumination setup was developed for a long-range imaging applications aiming at the detection of objects. A camera with a 300 mm lens and an aperture of 2.8 was chosen, which leads to a spatial resolution of the PMD chip of 1.5 cm in 100 meters distance. The spatial resolution of course limits the size and type of objects which can be detected. Moreover, it is obvious that standard lighting devices for PMD cameras will not work, since their range of application is limited to a few meters, i.e. 5 to 7 meters. Different types of setups have been tested and it was discovered that parabolic mirrors perform best, since a nearly parallel projection is required to reach distances over 100 meters. The required opening angle is about 1.5 degrees and here it seems intractable to achieve an effective and uniform lighting with collimators. The long-range lighting system incorporates four parabolic mirrors with corresponding LEDs which are carefully adjusted to cover the visible area of the camera. It is possible to orientate the system

Figure 3.15: Multi-modal imaging system: 2D/3D MultiCam, lighting devices and thermal camera.

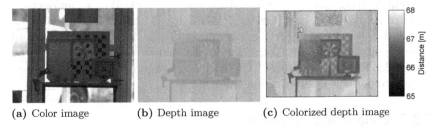

(a) Color image   (b) Depth image   (c) Colorized depth image

Figure 3.16: Indoor test of the multi-modal imaging system with a depth image evaluation setup in about 65 meters distance. A modulation frequency of 20 MHz and an exposure time of 10 ms were used.

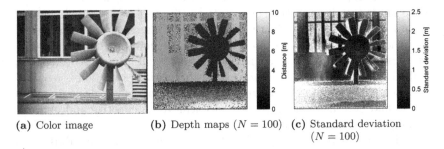

(a) Color image   (b) Depth maps ($N = 100$)   (c) Standard deviation ($N = 100$)

Figure 3.17: Outdoor test of the multi-modal imaging system with a scene in about 46 meters distance, which was acquired using a modulation frequency of 15 MHz and an exposure time of 10 ms.

horizontally and vertically. A stepping motor with appropriate controller is used for the vertical axis and the horizontal orientation is achieved with an external system. In Fig. 3.15 a development version of the multi-modal imaging system is depicted.

A series of evaluation tests has been performed, since the imaging system uses a novel lighting device and PMD based Time-of-Flight cameras have not been applied to these long distances before. In Fig. 3.16 some results of an indoor test are shown. Here a depth imaging test setup is used, which consists of checkerboard patterns, 3D Siemens stars and some common objects. This setup is located in 65 meters distance to the imaging system and a modulation frequency of 20 MHz with an exposure time of 10 ms was applied. The resulting depth and color images are of satisfactory quality besides the previously discussed intensity related distance error, which is

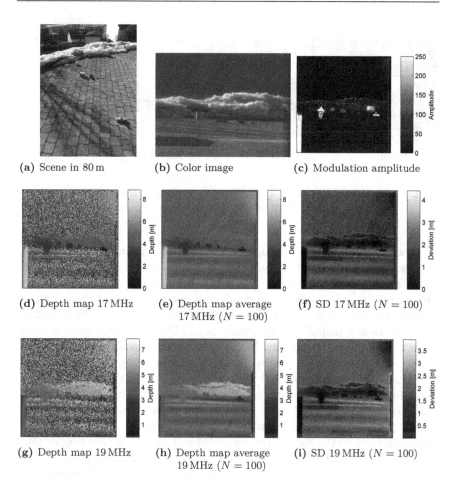

(a) Scene in 80 m

(b) Color image

(c) Modulation amplitude

(d) Depth map 17 MHz

(e) Depth map average 17 MHz ($N = 100$)

(f) SD 17 MHz ($N = 100$)

(g) Depth map 19 MHz

(h) Depth map average 19 MHz ($N = 100$)

(i) SD 19 MHz ($N = 100$)

Figure 3.18: Scene in 80 meters distance with depth maps, average of 100 depth maps and standard deviations acquired with 1 ms exposure time and 17 MHz.

noticeable in particular on the checkerboard pattern. Another medium distance test is shown in Fig. 3.17. The propeller is in 46 meters distance to the camera and the estimated standard deviation of the depth measurements is also displayed.

Results for a more challenging test are given in Fig. 3.18. The scene is positioned in 80 meters distance and was acquired on a sunny day in winter.

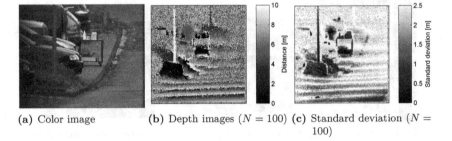

(a) Color image   (b) Depth images ($N = 100$) (c) Standard deviation ($N = 100$)

Figure 3.19: Images of a test acquisition in 130 meters distance using an integration time of 10 ms and a modulation frequency of 15 MHz.

An exposure time of 1 ms appropriate for outdoor imaging was chosen and 100 images were recorded for modulation frequencies of 17 and 19 MHz to estimate the standard deviation. The depth measurements for the test object are very precise, but the floor is difficult to measure because of the low angle of incidence. Results for a test setup in 130 meters distance are displayed in Fig. 3.19. Here a modulation frequency of 15 MHz and an exposure time of 10 ms were applied. The results display a significant amount of noise, which can be reduced largely by averaging or by applying longer exposure times. The experiment was performed with an earlier and less powerful version of the lighting device. The phase wraps are very prominent in the image showing the average, which verifies the correct distance measurements. The relatively small objects on the table can be recognized. Nevertheless, this distance is close to the limit of the capabilities of the 2D/3D imaging system.

Lastly results gathered in a field test are displayed in Fig. 3.20. Several objects were placed on an airfield in various distances. Larger objects can be recognized with a 300 mm lens even in 150 meter distance, but smaller object like screws could not be detected accurately at these distances because of the limited lateral resolution.

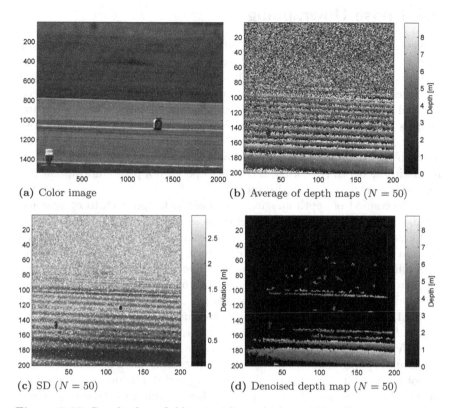

(a) Color image

(b) Average of depth maps ($N = 50$)

(c) SD ($N = 50$)

(d) Denoised depth map ($N = 50$)

Figure 3.20: Results for a field test with an object in 154 m distance. An exposure time of 5 ms and a modulation frequency of 17 MHz were applied.

3.4 Phase Unwrapping

Phase unwrapping to remove ambiguities in depth measurements is a common task in several research areas, e.g. remote sensing. Continuous wave amplitude modulation Time-of-Flight depth measurements do not measure distances directly, but derive the distance from the phase difference between emitted and received light. This means that the distance measured is ambiguous and multiples of the half wavelength $\lambda = \frac{c}{2\nu}$, with c being the speed of light, can be added. If the observed scene contains longer distances than λ and if the modulation frequency should not be decreased in order the to maintain the depth resolution, methods to perform a phase unwrapping are required to retrieve correct distances.

In the context of depth imaging, a probabilistic approach to remove ambiguities in a single depth image is proposed in [19]. An optimization is performed based on a cost function, which aims at the removal of discontinuities. In [18] this method was extended to incorporate multiple measurements with different modulation frequencies and hence different ambiguity ranges. A different approach based on a single depth map is introduced in [48]. The unambiguous depth of an object is inferred here by observing how much infrared light it reflects. The approach does not handle each pixel individually, but finds edges in the depth map in order to account for the different reflectivity of objects, which would otherwise compromise the results. The phase unwrapping methods are not completely stable and a method to obtain smooth results in subsequent phase unwrapped depth maps is described in [13]. Moreover, in [12] phase unwrapping is performed based on stereo information. The approach discussed in the next section was also published in [L10].

3.4.1 Multi-Frequency Approach and Results

Let $d_1, d_2, \ldots, d_n$ be distances measured with modulation frequencies $\nu_1, \nu_2, \ldots, \nu_n$. Then the most likely absolute distance $\hat{d}_i = d_i + k_i \frac{c}{2\nu_i}$ of all measurements combined can be obtained by minimizing the squared difference of all proposed absolute distances

$$min_{k_1,\ldots,k_n \in \mathcal{N}_0} \left\{ \sum_{i=1}^{n-1} \sum_{j=i+1}^{n} \left(d_i + k_i \frac{c}{2\nu_i} - d_j - k_j \frac{c}{2\nu_j} \right)^2 \right\} . \quad (3.17)$$

Since the measurement noise is not equal for different frequencies, it may be advantageous to weight the differences in the minimization with the variance of measurements at different frequencies. Using the mean of the measurements simplifies the minimization greatly if several images are taken per frequency. Let $\nu_1, \nu_2, \ldots, \nu_m$ with $m \leq n$ be the applied frequencies and let μ_u be the mean and σ_u^2 the variance of all measurements taken at frequency ν_u. Now the weighted minimization is given by

$$
min_{k_1,\ldots,k_m \in \mathcal{N}_0} \left\{ \sum_{u=1}^{m-1} \frac{1}{\sigma_u^2} \sum_{v=u+1}^{m} \frac{1}{\sigma_v^2} \left(\mu_u + k_u \frac{c}{2\nu_u} - \mu_v - k_v \frac{c}{2\nu_v} \right)^2 \right\}.
$$

(3.18)

The obvious approach is to combine a small frequency (e.g. 1 MHz with an ambiguity range of 150 m) and a larger frequency with a high depth resolution. However, this poses a severe challenge to the lighting devices and may lead to an unstable behavior or even damages. Moreover, the depth measurements are subject to a constant measurement error, which is specific to the modulation frequency as well as the lighting device. This offset needs to be determined for each frequency, otherwise wrong real distances are obtained, see [43] and [46] for PMD calibration methods. Therefore, we use just a small set of similar frequencies, which also allows to perform the optimization just calculating all possibilities (up to a given maximum distance).

The following experiment was conducted to illustrate the measurement characteristics for different modulation frequencies and their influence on the unwrapping results. In Fig. 3.21 images of a test setup were acquired

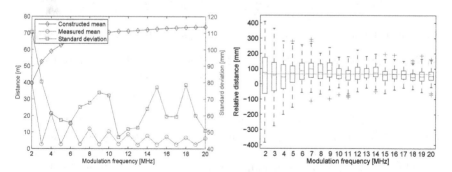

Figure 3.21: PMD measurements for different modulation frequencies. Left: mean and variance of the depth measurements as well as the constructed distance. Right: box plot of measurements of a relative distance of 6.5 cm (horizontal line).

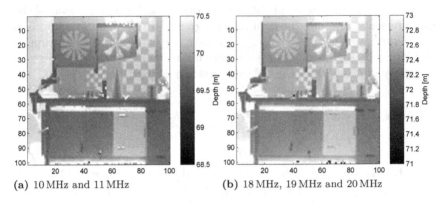

(a) 10 MHz and 11 MHz   (b) 18 MHz, 19 MHz and 20 MHz

Figure 3.22: Multi-frequency combination to eliminate ambiguities. 10 images per frequency were averaged and the real distance is 65 meters.

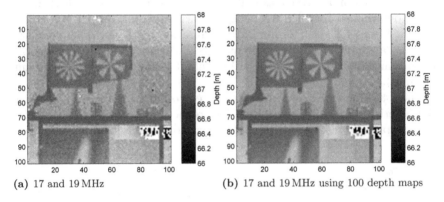

(a) 17 and 19 MHz   (b) 17 and 19 MHz using 100 depth maps

Figure 3.23: Enlarged results for another phase unwrapping indoor test using two frequencies and a low exposure time of 1 ms.

at different modulation frequencies. 100 images were taken per frequency and the mean and variance were computed for a chosen pixel. Additionally, measurement ambiguities were compensated for with the help of prior knowledge. As expected a declining variance can be observed for higher frequencies. Additionally, a box plot describing measurements of the relative distance between two objects in the scene is display in the figure.

In Fig. 3.22 two reconstructions using different frequencies are depicted. The setup is placed in 65 meters distance and 10 images were recorded

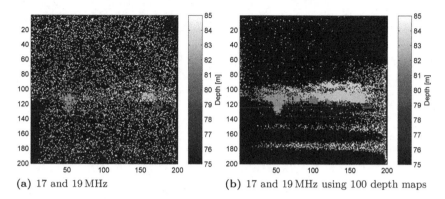

(a) 17 and 19 MHz (b) 17 and 19 MHz using 100 depth maps

Figure 3.24: Real world outdoor test again using two modulation frequencies and an exposure time of 1 ms.

per frequency. On the left 10 and 11 MHz were applied and on the right 18, 19 and 20 MHz. The reconstruction took only fractions of a second and it is accurate (only the constant distance measurement error is not eliminated here). For real-time imaging these calculations may be performed only once and afterwards only for regions with significant change or unlikely combinations of frequencies may be pruned, e.g. if two frequencies are used, only two possibilities have to be computed for one interval k_1 of the first frequency. Another approach to deal with invalid measurements and to speed up computations is to incorporate information of neighbor pixels.

A similar test setup was acquired with a low exposure time of 1 ms to emulate outdoor imaging and two modulation frequencies of 17 and 19 MHz were applied. In Fig. 3.23 enlarged areas of the phase unwrapped results are displayed, once using just two images and once using 100 depth maps per frequency. Results for a real world test to detect objects in 80 meters distance using the same parameters are shown in Fig. 3.24, see also Fig. 3.18. The phase unwrapping works reliably in these distances even when based on just two single depth maps.

4 Calibration of Depth Cameras

In recent years interest in 3D imaging gained considerable momentum and today 3D imaging is utilized in many different applications. Several techniques capable of providing depth measurements at video frame rate and with reasonable resolution are on the market. These devices are based on structured light (SL), on the Time-of-Flight principle or utilize stereo vision. All of these cameras are used in a wide range of applications. In some areas accurate depth measurements are not that important, i.e. in some video games or video chats range cameras are only employed to perform a fast and reliable foreground segmentation. However, when an application aims at the reconstruction of the observed scene, e.g. in order to maneuver in the scene or to correlate the observations of several cameras, the depth measurements delivered by any range camera need to be as accurate as possible.

Nevertheless, range cameras are typically subject to systematic measurement errors, which in turn make a calibration of the measurements necessary. Time-of-Flight cameras usually have a significant constant measurement error due to signal propagation delays. Additionally, errors caused by different light intensities and the actual distance to the reflecting object occur, see Section 3.1. Basically all range cameras are calibrated in order to account for deterministic measurement errors. The calibration usually consists of taking a number of images of a calibration object or pattern in known distances and then to use these to compensate for the measurement errors. The distances should cover the whole measuring range. Depth values not being included in the set of images taken are interpolated. Due to the fact that this procedure is relatively complex, the calibration is usually only performed once. This poses the question whether the calibration is accurate for the complete lifetime of the camera and for all imaging parameters such as the modulation frequency of continuous wave ToF cameras or the level of ambient light. Obviously, the calibration can only be valid for one type of illumination system.

Therefore, self-calibration approaches for range cameras have been developed as an extension of self-calibration techniques for normal video cameras. The methods operate with a number of images of a calibration object, e.g. a checkerboard, without known distances to the camera. Features of the calibration object are automatically detected and an optimization algorithm then determines the most likely camera parameters based on 2D and 3D point correspondences. These approaches are significantly less accurate and it is very difficult to cover the whole measuring range. Methods for depth camera calibration will be discussed further in Section 4.1.

Taking this one step further so-called auto-calibration methods are able to derive camera parameters without the need for any calibration objects and without known distances. These methods are extremely useful if the camera calibration is required but just a video or a single image is available (and the camera as well as its configuration are unavailable). Additionally, if the calibration is needed quickly or if the camera parameters are frequently changed these auto-calibration methods can be applied. Certain assumption regarding image features such as parallel lines or planes are utilized in these methods. Auto-calibration methods capable of calibrating the distance error for range cameras have not been published previously.

In this chapter several methods to handle and partly overcome depth measurement errors are presented. Methods from other authors will be discussed when relevant. Firstly, in Section 4.2 a method to compensate the varying efficiency of different PMD pixels (fixed pattern noise) is introduced. This calibration is necessary in order to make use of the PMD grayscale image or when phase measurements of different pixels shall be combined, e.g. for motion compensation. Afterwards a depth camera model is introduced in Section 4.3. At the end of this chapter a method to derive camera parameters from a single arbitrary depth image is detailed in Section 4.4.

4.1 Literature on Depth Camera Calibration

Camera calibration is a widely studied research topic. The book "Multiple View Geometry in Computer Vision" [27] gives an excellent overview and serves as a good starting point. In particular, self-calibration and auto-calibration of standard cameras, i.e. the estimation of camera parameters without special calibration objects or recording procedures, is covered. Therefore, only papers investigating the calibration of depth cameras will be reviewed in the following.

In the past, methods to calibrate depth cameras have been published in several papers, i.e. methods to perform a standard camera calibration and a distance calibration in order to account for measurement errors introduced in Section 3.1. Typically distance error models are introduced and their parameters are estimated. The distance error is then compensated with these models in order to gain accurate distance measurements. In [41] B-Splines were used to model the intensity related distance error, and the standard intrinsic camera parameters were determined conventionally using image features and bundle adjustment. This work was revisited and refined in [43] and in [58] the distance calibration was fused with the standard intrinsic parameter estimation and a polynomial distance error model was utilized. Comparable approaches were presented in [22, 39, 11]. An elaborate calibration device is used in [34] to compensate for different intensities due to different exposure times with a look-up table, which were also applied for this purpose in [53].

Multi-path reflections of the light emitted by ToF cameras is another error source for ToF distance measurements, which is discussed and possibly compensated for in [20]. Moreover, the combination of multiple ToF cameras and the effects of operating multiple ToF cameras at the same time is explored in [37].

4.2 Intra-Pixel Intensity Calibration

Pixel values of a grayscale image of a PMD camera must be comparable when these images are utilized in a given task. However, this requires an intensity calibration of all PMD pixels to account for the so-called fixed pattern noise. The same calibration is necessary, when camera motions are supposed to be compensated for by combining shifted phase images, see Section 6.2. In Fig. 4.1 the raw intensity values for one channel of a phase image are shown. The PMD imaging chip utilizes three AD converters with different effective offsets and gains, which results in vertical lines to pixel variations. In order to perform the normalization, a set of images of a uniformly lit diffuse glass is acquired at different exposure times. The relative offsets and gains for both channels of each pixel can be estimated based on these images assuming linearity with a linear regression, which enables an affine normalization of the phase images. The resulting intensity values for one channel are also given in Fig. 4.1. An earlier approach in [40] aims at reducing inhomogeneities of PMD grayscale image, but it includes the modulated light in the intensity calibration and uses planes of different reflectivity instead of varying the exposure time to generate different gray levels. Since the amount of active light observed at a certain pixel depends non-linearly on the distance to the object and since the usage of only a few different planes is feasible, the proposed approach promises to be superior. Nevertheless, it does not

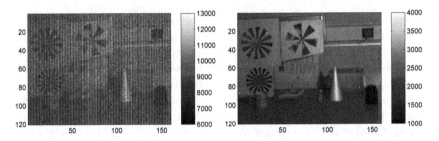

Figure 4.1: Affine normalization of a PMD phase image. Left: original phase
image for the first PMD channel, right: normalized image.

account for global differences in illumination, because it only calibrates the
PMD chip.

The exact calibration procedure is detailed in the following. Let Γ^i for
$\Gamma \in [A, B]$ and $i = 1, \ldots, n$ be n phase images for a PMD channel Γ.
The index i describes here different acquisitions, since the phase is not of
importance now. The pixels of the image are denoted by $\Gamma^i_{(x,y)}$ and let t_i be
the associated integration time. Then the average pixel value $\mu^\Gamma_{(x,y)}$ is given
by

$$\mu^\Gamma_{(x,y)} = \frac{1}{n} \sum_{i=1}^{n} \Gamma^i_{(x,y)} \ . \tag{4.1}$$

The affine parameters for each channel $\alpha^\Gamma_{(x,y)}$ and $\beta^\Gamma_{(x,y)}$ are calculated with

$$\tau_i = t_i - \frac{1}{n} \sum_{j=1}^{n} t_j \tag{4.2}$$

$$\alpha^\Gamma_{(x,y)} = \frac{\sum_{i=1}^{n} \tau_i \cdot \left(\Gamma^i_{(x,y)} - \mu^\Gamma_{(x,y)} \right)}{\sum_{i=1}^{n} \tau_i^2} \tag{4.3}$$

$$\beta^\Gamma_{(x,y)} = \mu^\Gamma_{(x,y)} - \alpha^\Gamma_{(x,y)} \frac{1}{n} \left(\sum_{j=1}^{n} t_j \right) \ . \tag{4.4}$$

Now the normalized pixel value $\hat{\Gamma}_{(x,y)}$ of a PMD channel Γ for an acquired pixel value $\Gamma_{(x,y)}$ can be computed with

$$\mu_\alpha^\Gamma = \frac{1}{|\Gamma|} \sum_{(x,y)\in\Gamma} \alpha_{(x,y)}^\Gamma \tag{4.5}$$

$$\hat{\Gamma}_{(x,y)} = \left(\Gamma_{(x,y)} - \beta_{(x,y)}^\Gamma\right) \cdot \frac{\mu_\alpha^\Gamma}{\alpha_{(x,y)}^\Gamma} . \tag{4.6}$$

4.3 Depth Camera Model

In the following a pinhole camera model for range cameras is described in detail. The camera model maps a pixel and a depth measurement to a 3D point. The parameters of this model and the notations are summarized in Table 4.1. A pixel (x,y) is mapped to metric 2D coordinates (x_c, y_c) in relation to the principal point (c_x, c_y) on the image plane with

$$x_c = (x - c_x)w_p \tag{4.7}$$

$$y_c = (y - c_y)h_p \tag{4.8}$$

for a pixel width w_p and height h_p. The z-coordinate of the 3D point (X, Y, Z) with the $X - Y$ plane being identical to the image plane can be calculated based on these 2D coordinates and the distance $\delta(d)$ to the principle point with

$$Z = \frac{\delta(d)}{\sqrt{1 + \frac{(x_c^2 + y_c^2)}{f^2}}} \tag{4.9}$$

and this leads to the x- and y-coordinates

$$X = \frac{x_c Z}{f} \tag{4.10}$$

$$Y = \frac{y_c Z}{f} \tag{4.11}$$

with the focal length f. Range cameras often do not provide depth measurements directly or provide only measurements which need to be corrected. In this range camera model a function $\delta(d)$ maps a raw depth measurement to a depth in meters. For PMD based cameras, which utilize phase shift information, a simple distance function is given by

$$\delta(d) = \frac{c(2\pi d + o)}{4\pi\nu} \tag{4.12}$$

Table 4.1: Camera Parameters and Notations.

Symbol	Quantity	Meaning
x, y	pixel	Coordinates of the pixel observing the object
d	$\in [0, 1]$	Raw depth measurements
w_p, h_p	meter	Width and height of a pixel
c_x, c_y	pixel	Position of the optical center on the imaging chip
$\delta(\cdot)$	meter	Function mapping a raw depth measurement to the distance of the object
ν	Hz	Modulation frequency
o	rad	Constant distance error (phase offset)
x_c, y_c	meter	2D coordinates of the pixel on the image plane
f	meter	Focal length of the camera
X, Y, Z	meter	3D camera coordinates

with c being the speed of light, ν the modulation frequency used and o a constant measurement offset. Errors of higher order were modeled in the past, cf. [46], [34], [53] or [43]. In Section 3.1 it was shown that the measurement error mainly depends on the phase shift and the observed intensity, which depends on the active lighting as well as the ambient light. Therefore, if higher order models are applied, they should depend on phase, amplitude modulation and intensity or the calibration will be specific to the scene, i.e. level of ambient light.

4.4 Auto-Calibration

It can be seen from camera model equations 4.9-4.12 that the 3D coordinates X, Y, Z are nonlinear in all intrinsic camera parameters c_x, c_y, f, o with the frequency ν being an exception. Therefore, deviating from the true parameters in the reconstruction of the scene geometry will result in deformed objects, e.g. planes display a significant curvature and angles between objects get modified as illustrated in Fig. 4.2a. It was discovered in the experiments that this property can in fact be exploited to reliably estimate the constant distance error.

Since many man-made objects are planar (walls, floors, tables), their reconstruction can be utilized to find the correct parameters by refining them until the object in question is actually planar. In theory all non-linear parameters can be estimated using a single plane in the scene.

The proposed auto-calibration method for depth cameras consists of several processing steps. Firstly, a segmentation of surface normals in 3D is performed to detect large planes in the scene. Based on the detected planes, a novel auto-calibration approach is able to determine the distance offset and the focal length of a depth camera. The capabilities of this approach, especially its accuracy, as well as limitations and possible applications of this approach are discussed in the following.

4.4.1 Segmentation of Surface Normals

In order to recognize planes in the scene a segmentation or feature space clustering of surface normals in 3D is performed. The normal vectors are derived from a depth map and the segmentation is carried out with the help of the Mean-Shift algorithm as described in [14]. The depth values for each pixel are initially mapped to 3D points with the range camera model and a set of default camera parameters. Then the surface normal $\underline{n} = (n_x, n_y, n_z)$ at each 3D point $\underline{p} = (p_x, p_y, p_z)$ can be estimated by averaging over the eight normal vectors of the triangles spanned by $\underline{p}$ and its neighboring points. This leads to interpolation errors at the borders of objects and therefore boundary points of detected planes have to be neglected in the auto-calibration later on.

The Mean-Shift algorithm is a feature space approach and consists of two steps. In the first step (filtering) the mean-shift vectors are calculated iteratively and the feature points are moved accordingly until a convergence is reached. These vectors are determined by calculating the weighted average of neighboring feature points. Let $P_i = (\underline{p}_i, \underline{n}_i)$ and $P_j = (\underline{p}_j, \underline{n}_j)$ be two points of the feature space. The weight $g(P_i, P_j)$ between the two points is constructed as the product of two Gaussian kernels

$$g(P_i, P_j) = \exp\left\{ -\frac{\left(\underline{p}_i - \underline{p}_j\right)^T \left(\underline{p}_i - \underline{p}_j\right)}{\sigma_{space}^2} \right\} \cdot \exp\left\{ -\frac{1 - \left(\frac{\underline{n}_i \cdot \underline{n}_j}{\|\underline{n}_i\| \|\underline{n}_j\|}\right)^2}{\sigma_{angle}^2} \right\} .$$

(4.13)

The first term is based on the squared Euclidean distance between the two points and the second uses the squared sine to measure differences between normal vectors. σ_{space} and σ_{angle} are bandwidth parameters to control the influence of the different subspace measures.

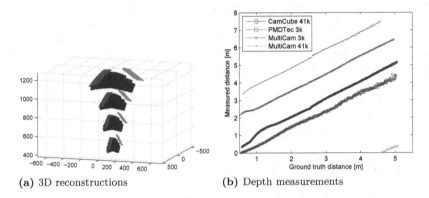

(a) 3D reconstructions   (b) Depth measurements

Figure 4.2: 3D reconstructions of a simple scene consisting of three orthogonal walls while applying different offsets and uncorrected range
measurements of four different ToF cameras

Given a point $P_i^{(t)}$ in the feature space at iteration t, the corresponding
point in the next iteration can be computed with

$$P_i^{(t+1)} = \frac{\sum_{P \in N(P_i^{(t)})} P \cdot g(P_i^{(t)}, P)}{\sum_{P \in N(P_i^{(t)})} g(P_i^{(t)}, P)} , \qquad (4.14)$$

where the set $N(\cdot)$ describes a spatial neighborhood. The Mean-Shift vector
is the difference $P_i^{(t+1)} - P_i^{(t)}$ and these iterative calculations are performed
for each point independently until convergence is observed. It can be proven
that the series always converges for Gaussian kernels, cf. [14]. Once the
Mean-Shift filtering is finished, the feature points are merged using a distance
measure in the feature space and a threshold in order to form the segments.

4.4.2 Proposed Estimation of Camera Parameters

Uncorrected depth measurements for four different range cameras all based
on PMD chips are depicted in Fig. 4.2b. These measurements were conducted
with the help of a precise linear translation unit.

The measurement errors can roughly be classified as follows: constant
distance errors, errors due to over-exposure or insufficient lighting and an

error of higher order sometimes called wobbling. Obviously the constant distance error or phase offset needs to be corrected, whereas the exposure related errors cannot be compensated for and the higher order errors are minor. Moreover, it cannot be expected to compensate for these errors without elaborate measurements and thus the simple depth camera model described in Section 4.3 will used in the following.

Related to the normal depth calibration of range cameras is the validation of range measurements. Small changes to the camera setup, e.g. different cables or lighting devices used in conjunction with PMD based cameras, lead to changes of the phase offset, but also applying different modulation frequencies as is demonstrated in Fig. 4.6. Aging or wear may also impair depth measurements as does the relative positioning of lighting devices and camera. This is also a concern for camera stereo setups.

The proposed auto-calibration relies on the existence and successful detection of at least one plane of significant size. The segmentation of surface normals is conducted with an initial set of camera parameters to retrieve candidate planes in the scene by selecting the largest segments, which additionally need to be larger than a given threshold. Then a plane is fitted to each candidate plane in the scene. This could be achieved with a simple least squares method, but instead a RANSAC based algorithm (as described in [21]) is applied, which itself is based on the well known eigenvalue based plane fitting method. This algorithm is capable of handling outliers, e.g. at borders of objects, and thus yields good results.

Now a measure $\rho_{curv}(\cdot)$ is utilized to judge the curvature of a plane and thereby the accuracy of the camera parameters. $\rho_{curv}(\cdot)$ is simply given by the square root of the average squared distance between all points of the plane to the constructed plane. Let $\underline{n}_1, \underline{n}_2, \ldots, \underline{n}_m$ be the normal vectors of unit length of the m constructed planes with points on the plane $\underline{p}_1, \underline{p}_2, \ldots, \underline{p}_m$ and let $M_1, M_2, \ldots, M_m$ be sets of the 3D points used in the construction. The curvature measure can now be defined as

$$\rho_{curv}(M., \underline{n}., \underline{p}.) = \frac{1}{m} \sum_{i=1}^{m} \sqrt{\sum_{\underline{q} \in M_i} \left(\left(\underline{q} - \underline{p}_i \right) \cdot \underline{n}_i \right)^2}. \qquad (4.15)$$

A second measure to judge the camera parameters applied in the reconstruction is based on the angles between the detected planes. Here at least two pairwise orthogonal planes of significant size have to be detected successfully in the scene. Then a measure to rate the orthogonality of the detected planes in the scene is given by

$$\rho_{ortho}(\underline{n}_1, \underline{n}_2, \dots, \underline{n}_m) = \frac{2}{m(m-1)} \sum_{i=1}^{m-1} \sum_{j=i+1}^{m} \left(\frac{\underline{n}_i \cdot \underline{n}_j}{\|\underline{n}_i\| \|\underline{n}_j\|} \right)^2. \qquad (4.16)$$

The function $\rho_{ortho}(\cdot)$ is an average of the squared cosines of the angles between the normal vectors of the planes. Now a standard non-linear optimization method such as the Levenberg-Marquardt algorithm can be applied to estimate camera parameters.

4.4.3 Experimental Evaluation

In Fig. 4.3 exemplary results for the Mean-Shift segmentation of surface normals extracted from a range image are displayed. A set of different scenes

(a) Grayscale   (b) Depth   (c) Response of the curvature
                                                   measure

(d) Segm. $\sigma_{angle} = 10$ (e) Segm. $\sigma_{angle} = 15$ (f) Segm. $\sigma_{angle} = 20$

Figure 4.3: Mean-Shift segmentation based on surface normals of a real world scene for different bandwidth parameters.

and lenses were tested with different bandwidth parameters to confirm the robustness of the segmentation and a good-natured behavior was observed, which is by far superior to a segmentation based just on color or intensity. Results for different values of σ_{angle} with normal vectors scaled to length 100 and a spatial bandwidth of $\sigma_{space} = 5$ cm are shown.

In order to evaluate the curvature measure $\rho_{curv}(\cdot)$ a simple scene consisting of just a single wall was acquired with the PMDTec CamCube 41k and a set of different lenses. An area of 100×65 pixels was selected and the curvature measure was computed for reconstructions based on different

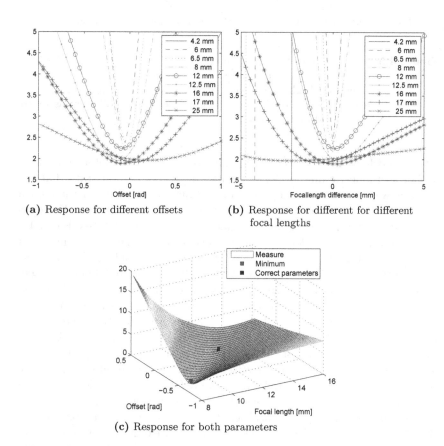

(a) Response for different offsets

(b) Response for different for different focal lengths

(c) Response for both parameters

Figure 4.4: Curvature measure evaluated for a number of acquisitions with different lenses of a single wall with a size of 100×65 pixels.

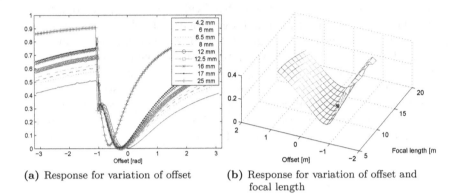

(a) Response for variation of offset   (b) Response for variation of offset and
                                                 focal length

Figure 4.5: Evaluation of the orthogonality measure. Left: acquisitions with
different focal lengths and variation of the offset used in the reconstruction. Right: single acquisition with a focal length of 12.5 mm
and evaluation of the orthogonality measure for different assumed
offsets and focal lengths

offsets and focal lengths. The true offset $o = -0.06$ was determined with
a photogrammetric method. The results in Fig. 4.4 demonstrate that the
offset as well as the focal length can be reliably estimated under these good
conditions if the other parameter is fixed to the correct value. However,
estimating both parameters at the same time is not possible (see Fig. 4.4c).
Additionally, large focal lengths cause problems due to the diminishing effect
of the offset. Fig. 4.3c shows the curvature measure applied to the largest
plane in the real world scene captured with a 12 mm lens and different offsets.

For the orthogonality measure a test scene consisting of three orthogonal
planes was captured again with lenses of different focal lengths. A simple
map describing which points belong to planes was created per hand. In
Fig. 4.5 results for the proposed orthogonality measure $\rho_{ortho}(\cdot)$ are displayed.
In Fig. 4.5 the parameter for the focal length in the reconstruction was set to
its correct value and the distance offset was varied. The results demonstrate
that the offset can be estimated quite reliably for smaller focal lengths which
are commonly used for scene observation applications. The accuracy of
the method deteriorates for focal lengths larger than 25 mm again due to
the decreasing effect of the offset. In order to examine the influence of
incorrect focal lengths on the 3D reconstruction the orthogonality measure
is calculated for an acquisition with a 12 mm lens and the results are plotted

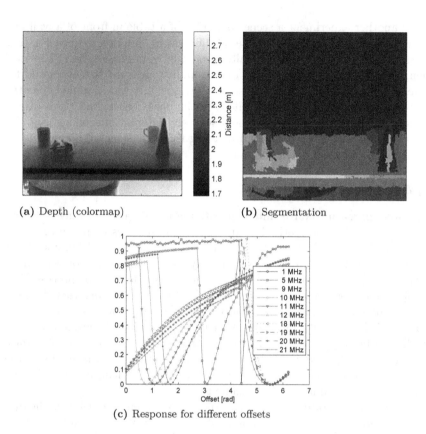

(a) Depth (colormap)   **(b)** Segmentation

(c) Response for different offsets

Figure 4.6: Acquisitions of a second test scene with different modulation frequencies. Segmentation of surface normals with the Mean-Shift algorithm and evaluation of the orthogonality measure for different distance offsets.

in Fig. 4.5b. The results demonstrate that the correct focal length cannot be estimated based on the proposed measure and both camera parameters cannot be estimated simultaneously. Therefore, standard methods based on the recognition of lines etc. are required for a complete depth camera auto-calibration. These results were confirmed with a scene containing two orthogonal walls, which was captured at larger distances (3 to 6 meters depending on the focal length) and with a real world scene (Fig. 4.3).

In another experiment a scene consisting of a table in front of a wall and some objects was acquired for a set of different modulation frequencies. The orthogonal walls were automatically recognized via segmentation. The focal length used was 17 mm and the orthogonality measure was evaluated for the whole range of potential offsets. Previously, it was assumed that the modulation frequency has little effect on the phase offset, but it was discovered that different frequencies yield severely different offsets, see Fig. 4.6. This was confirmed by visually inspecting the reconstructed planes and by applying photogrammetric methods.

4.4.4 Summary

A novel approach towards auto-calibration and validation of range cameras was presented. The method is applicable to all types of range or depth cameras, but it was tested with ToF cameras based on the Photonic Mixer Device here. In general, a scene can only be reconstructed correctly from depth images if the correct camera parameters are used in the process. The proposed approach make use of the fact that planes are reconstructed with a curvature and that angles between planes are modified if wrong parameters are applied. It was demonstrated that large planes like the ground plane or walls can be reliably detected with the Mean-Shift segmentation of surface normals. The geometry of these planes can then be used to estimate the constant distance measuring error as well as the focal length of the camera, but not both parameters at the same time.

Two measures to judge the reconstruction of a scene are defined: the first one measures how planar the points of a detected plane are and requires a single plane of significant size in the scene. For the second measure angles between detected planes are computed while assuming that the detected planes are orthogonal. This assumption is satisfied in many practical cases. It was not possible to estimate errors of higher order or to estimate camera parameters simultaneously. Nevertheless, other well-known auto-calibration methods also relying on the successful detection of scene geometry could be applied in conjunction with the proposed method for a complete auto-calibration of any range camera. Other possible applications for the proposed measures are a permanent validation the depth measurements of a camera or to estimate the focal length when a zoom lens is used and the focal length is changed often.

5 Multi-Modal Background Subtraction

Background subtraction is a common first step in the field of video processing and its purpose is to reduce the effective image size in subsequent processing steps by segmenting the mostly static background from the moving or changing foreground. In this chapter approaches towards background modeling previously described are extended to handle 2D/3D videos. The background is estimated using the widespread Gaussian mixture model for color as well as for depth and amplitude modulation information. A new matching function is presented, which allows a better treatment of shadows and noise and reduces block artifacts.

Limitations to overcome the problem of fusing high resolution color information with low resolution depth data are addressed and the approach is tested with different parameters on several scenes. The results are compared to standard and widely accepted methods.

5.1 Overview

This chapter focuses on the common image processing step of background subtraction, modeling or segmentation. The goal is to isolate regions of interest in each image, which are here defined as the moving foreground as opposed to the mostly static background of the video. Background subtraction methods can be divided roughly into two categories: pixel- and region-based approaches. The former perform the classification of a pixel based only on already known information about that pixel, whereas the latter use information from neighboring pixels (often grouped into regions or layers) as well. The most notable of the pixel-based approaches is the method of Gaussian Mixture Models, which is widespread, simple and efficient. The background is here modeled as a mixture of Gaussians for each pixel - hence the name. When the observations are accompanied by a depth value and an amplitude modulation for each pixel, the problem has to be addressed how to combine these different types of dimensions in the classification step.

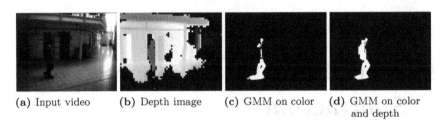

(a) Input video   (b) Depth image   (c) GMM on color   (d) GMM on color
                                                                 and depth

Figure 5.1: Exemplary results for (a) a challenging input video , (b) a depth
image where invalid measurements are black, (c) the foreground
mask using GMM only for color and (d) GMM based on color and
depth.

In this work the background subtraction method based on Gaussian
Mixture Models (GMM) is adapted to videos with color, depth and amplitude
modulation information gathered through the Time-Of-Flight principle,
which will be referred to as 2D/3D videos (see Fig. 5.1 for an example). Here
the significantly lower resolutions of the depth and amplitude modulation
images have to be accounted for additionally. To that end a measure is
presented that links the dimensions in a statistical manner, which also results
in a lower noise level and a better classification of shadows. Previous methods
either use rather simple foreground estimation methods or are designed to
operate on full-sized depth images which were acquired with a stereo setup.
The approach was also published in [L1, L3].

5.2 Related Work

In [24] the foreground of 2D/3D videos is simply extracted by defining a
volume of interest in 3D space and these foreground maps are exploited for
hand tracking as well as gesture recognition. Harville et al. applied the
standard approach of background modeling by Gaussian mixtures, see e.g.
[60], to color and depth videos in [28]. They operate on full-sized depth
images so that there is no need to handle different resolutions.

In [6] a rather simple approach for foreground segmentation for 2D/3D
videos is evaluated, which is based on region growing and refrains from
modeling the background, whereas in [38] a simple pixel-based background
modeling method called ViBe is employed for color and depth dimensions

separately. The resulting foreground masks are fused with the help of binary image operations such as erosion and dilation.

A more elaborate method of fusing color and depth information is bilateral filtering (see Section 6.1.2), which is used e.g. in [15]. Here a preliminary foreground is approximated with the help of a dividing plane in space and a bilateral filter is applied to obtain the final results. The method is demonstrated on depth augmented alpha matting, which is also the focus of [65]. In [59] the ability of bilateral filtering to deal with geometric objects is demonstrated and in [10] a variant of the method designed to handle noise and invalid measurements is presented.

5.3 Multi-Modal Gaussian Mixture Models

All observations at each pixel position $\underline{x} = [x, y]^T$ are modeled with a mixture of Gaussians to estimate the background of a video seqeunce. The assumption is that an object in the line of sight associated with a certain pixel produces a Gaussian formed observation (or several in the case of a periodically changing appearance, e.g. moving leaves, monitor flickering). Each observation is then modeled with one Gaussian, whose mean and variance is adapted over time. An observation at time t for a pixel $\underline{x}$ is given by $\underline{s}^t(\underline{x}) = [s_1^t(\underline{x}), s_2^t(\underline{x}), \ldots, s_n^t(\underline{x})]^T$. The probability distribution density of $\underline{s}^t(\underline{x})$ can now be described by

$$f_{\underline{s}^t(\underline{x})}(\underline{\xi}) = \sum_i \omega_i \cdot N_{\underline{\xi}}\left(\underline{\mu}_i, \Sigma_i\right) , \tag{5.1}$$

where

$$N_{\underline{\xi}}\left(\underline{\mu}_i, \Sigma_i\right) = \frac{1}{(2\pi)^{\frac{\dim(s)}{2}} |\Sigma_i|^{\frac{1}{2}}} \exp\left\{-\frac{1}{2}\left[\underline{\xi} - \underline{\mu}_i\right]^T \cdot \Sigma_i^{-1} \cdot \left[\underline{\xi} - \underline{\mu}_i\right]\right\} \tag{5.2}$$

is the multivariate Gaussian with mean $\underline{\mu}_i$ and covariance matrix Σ_i. The mixing coefficients ω_i must satisfy $\sum_i \omega_i = 1$. The following questions concerning this model immediately come to mind: How many Gaussians should be employed to model the observations, how are the Gaussian adapted efficiently over time and which Gaussians should be considered background. Most GMM-based methods follow simple assumptions for efficiency reasons. They use a fixed number of Gaussians per pixel and the minimum number of Gaussians with weights, which sum up to a given threshold are treated as background.

The adaptation of the Gaussians over time is a bit more complicated. A simply online clustering approach is used in this work to estimate the parameters of the mixture instead of using the Expectation-Maximization (EM) algorithm or similar methods: It is checked for each new observation $\underline{s}(\underline{x})$ if it is similar to already modeled observations or if it is originating from a new object. It may also just be noise. This test consists of evaluating the Mahalanobis distance $\delta(\cdot, \cdot)$ to the associated Gaussian $N_{\underline{x}}(\underline{\mu}_i, \Sigma_i)$ with

$$\delta\left(\underline{x}, \underline{\mu}_i\right) = \sqrt{\left(\underline{\mu}_i - \underline{x}\right)^T \Sigma_i^{-1} \left(\underline{\mu}_i - \underline{x}\right)} < T_{near} \,, \qquad (5.3)$$

where T_{near} is a given threshold. When similar observations have been recorded, their Gaussian is adapted using the observed data. Otherwise, a new Gaussian is created and added to the mixture. An exact description of a possible implementation can be found in [60] for normal videos and in [28] with additional depth values.

An observation for a pixel is denoted by $\underline{s}(\underline{x}) = (y, c_b, c_r, z, a)^T$ in this thesis and contains the color value in YCbCr format, a depth value z and an amplitude modulation value a. The 2D/3D camera records a full-sized color image and low resolution depth and amplitude modulation images, which are resized to match to color images by the nearest neighbor method. The variances of all Gaussians are limited to be diagonal in order to simplify computations.

When working with ToF data, invalid depth measurements due to low reflectance, have to be handled cautiously. A depth measurement is considered invalid if the corresponding amplitude is lower that a given threshold. In [28] an elaborate logical condition is used to classify a new observation. Empirical results show that this can be simplified by using the measure

$$\widehat{\delta}\left(\underline{x}, \underline{\mu}_i\right)^2 = \left(\underline{\mu}_i - \underline{x}\right)^T \Sigma_i^{-1} \begin{pmatrix} 1 & & & \\ & \lambda_c & & \\ & & \lambda_c & \\ & & & \lambda_z & \\ & & & & \lambda_a \end{pmatrix} \left(\underline{\mu}_i - \underline{x}\right) \qquad (5.4)$$

and checking the condition

$$\widehat{\delta}\left(\underline{x}, \underline{\mu}_i\right)^2 < T_{near}^2 \cdot Tr \begin{pmatrix} 1 & & & \\ & \lambda_c & & \\ & & \lambda_c & \\ & & & \lambda_z & \\ & & & & \lambda_a \end{pmatrix} = T_{near}^2 \left(1 + 2\lambda_c + \lambda_z + \lambda_a\right) \,,$$
$$(5.5)$$

where $\lambda_z \in \{0, 1\}$, depending on whether current and previous depth measurements are both valid. The mechanism from [28] works well to that end.

Similarly, $\lambda_c \in \{0,1\}$ indicates whether the chromaticity channels of the current observation as well as the recorded information provide trustworthy values. This can be estimated by simply checking if both luminance values or their mean respectively are above a certain threshold. Finally, $\lambda_a \in \{0,1\}$ determines if the amplitude modulation should be used for the classification and is specified a priori.

This matching function take the fact into account that observations in the color, depth and amplitude modulation dimensions are in practice not independent. A foreground object has probably not only a different depth but also at least a slightly different color and infrared reflectance properties. Other reasons for the dependency are limitations of video and ToF cameras, e.g. the infrared reflectance of an object has an influence on the depth measurement (or its noise level) and low luminance impairs chromaticity measurements. Therefore, a linkage between the dimensions reduces the noise level in the foreground mask as well as the amount of mis-classification due to shadows and block artifacts, which occur when only the low resolution depth measurements differ.

More elaborate variations such as learning modulation and special treatment of deeper observations, when determining what observations are considered background, are described in [28] but seem to be unnecessary for simple scenarios.

5.4 Experiments

The approach described in this thesis can be evaluated by examining if the following objectives are achieved:

- When ordinary background subtraction based on color only works, then the additional information should not deteriorate the results, e.g. by introducing block artifacts at the border of the foreground mask.

- When the foreground is not classified correctly based on color, then this should be compensated by depth information.

- The treatment of shadows of the color based background subtraction is far from perfect and should be improved with depth information.

The following methods were compared in the course of this work. 'GMM' stands for the standard color-based GMM approach (see [60]) and 'Original GMMD' is the original color and depth-based method of [28]. 'GMMD without depth' denotes the method described in this thesis without considering

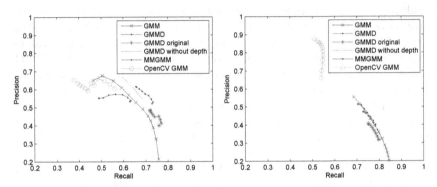

Figure 5.2: Average recall and precision values for different background subtraction methods using different parameters. Left: values for the
2D/3D video shown in table 5.1, right: for the video shown in
Table 5.2.

depth measurements (force $\lambda_z = 0$) and with $\lambda_a = 0$, whereas in 'GMMD'
λ_z is determined based on the amplitude modulation for each pixel similar
as described in [28] and in 'MMGMM' $\lambda_a = 1$ is set additionally. The values
for the OpenCV GMM method are given for reference only, since it includes
post-processing steps and is therefore not directly comparable.

In Table 5.1 the results of the background subtraction methods for a
2D/3D video taken under difficult lighting conditions are shown. The same
parameters were applied for all methods: a maximum number of 4 Gaussians
per pixel, a learning rate of $\alpha = 0.0005$, an initial $\sigma = 5$ and a threshold
$T_{near} = 3.5$. Due to the fact that all methods operate based on the same
principle, the results should be comparable for a given set of parameters.
This was also confirmed by testing several parameter variations.

The results demonstrate the ability of the proposed method to achieve
the mentioned objectives. The mis-classification of shadows is reduced and
the natural borders of the foreground are detected more accurately. When
the classification based on color fails, then the missing areas are filled at
least partly. Unfortunately, the compensation is often done in a block-wise
fashion (see also Fig. 5.1). This drawback is discussed further in the next
section.

Image sequences from another experiment are evaluated in Table 5.2 with
the same parameter set. Here the lighting conditions are far better, so that
the standard GMM algorithm can in theory distinguish between foreground

and background. However, shadows and similarities between the foreground (jacket) and the background cause large problems in this video. The proposed method does not affect the accurate classification based on color but allows for better shadow treatment due to the available depth values.

In Fig. 5.2 quantitative results for both 2D/3D videos are displayed. A ground truth was created by hand for every 5th frame starting with the last empty frame before the person enters the scene and ending with with first empty frame after the person has left the scene. Then the number of true positives tp, false positives fp and false negatives fn is counted in each frame for the different methods using different thresholds $T_{near} = 2, 2.5, \ldots, 8$ to calculate the recall $tp/(tp + fn)$ and the precision $tp/(tp + fp)$ values. Then their average over all frames is plotted. Here all variants of the proposed methods perform superior to the classic approach and to the original GMMD method with the exception of the MMGMM method when applied to the first video. However, this method achieves the best results for the second video. The poor results in respect to the first video are caused by the fact that the scene in this video is much more difficult to illuminate than the scene in the second video, which results in higher noise levels for the amplitude modulation images in the first video. The different values for the OpenCV GMM method for the second video are caused by the correct classification of the TV, whereas all other methods fail in that respect. However, the comparably low recall values for the OpenCV GMM method, i.e. a largely incomplete true foreground possibly due to foreground background similarities, are worth mentioning.

Table 5.1: Walk-by video with difficult lighting conditions.

Method	Frame 195	Frame 210	Frame 225	Frame 240	Frame 255
Input					
Depth					
Modulation amplitude					
GMM					
Original GMMD					
GMMD without depth					
GMMD MMGMMD					
OpenCV GMM					

Table 5.2: Simple video with good lighting conditions but difficult foreground.

5.5 Limitations

Due to its pixel-based nature the ordinary color-based GMM background subtraction cannot distinguish between foreground and background, when the color difference is small. The depth values gained from a ToF camera provide the opportunity to classify all image blocks correctly with depth values different from those of the background as long as there are valid depth measurements available for the background. As illustrated in Fig. 5.3 classification based only on low resolution depth values will result in an unnatural foreground contour due to block artifacts. This drawback cannot be resolved in all situations in practice, since the background usually continues with the same color in such areas, so that there is no edge that would allow gradient-based methods to smooth the contour of the foreground mask correctly. Otherwise bilateral filtering (see Section 6.1.2) which is often used in the context of 2D/3D videos to enhance the resolution of depth maps, would be able to reconstruct the true contour of the object.

In order to resolve the general case, contour estimation methods that incorporate knowledge of the object given a priori or learned through time are necessary, but it does not seem to be possible to achieve good results in a not strictly defined setting. In the opposite case, when a difference in depth measurements result in a wrong classification, gradient based methods can be applied to smooth the contour.

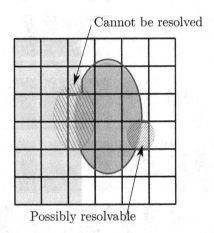

Figure 5.3: Illustration of different regions of a foreground mask. The foreground object, the background with similar color and the detected contour of an object are marked.

5.6 Summary

In this chapter the standard method for background subtraction based on Gaussian Mixture Models is adapted to operate on videos recorded with a 2D/3D camera. The method proposed was compared to standard as well as previous methods using simple 2D/3D video sequences. Qualitative as well as quantitative results were presented and it was demonstrated that the method proposed was able to compensate for mis-classifications of pixels due to color similarities between foreground objects and the background by utilizing depth and modulation amplitude information without harming the high resolution contour of foreground objects. Furthermore, this method provides a improved treatment of shadows and noise compared to methods previously published.

The additional computational demand to process and maintain the depth values is small compared to standard background subtraction methods based on GMM, i.e. on current PCs real-time processing is easily possible.

6 Super-Resolution for Depth Maps

One of the major limitations of depth cameras is the relatively low resolution, which is addressed in this chapter. The resolution of color images easily reaches mega-pixels in multi-camera 2D/3D setups or for the 2D/3D MultiCam (cf. Section 3.2). However, the resolution of current PMD chips is 160×120 or 200×200 pixels. Competitors also produce depth imaging chips with similar low resolutions. Depth imaging chips with higher resolutions from PMD Technologies and other companies exist, but they come with major limitations in range. In any case for certain applications high resolution depth maps generated from low resolution depth images are required, but a simple linear or quadratic scaling of the depth images will result in invalid depth values, whenever the real geometry of the scene differs from a planar or spherical one respectively, see Fig. 6.1 for an example.

In this chapter previously published super-resolution methods for 2D/3D images are detailed as well as evaluated in Section 6.1. These methods serve as a pre-processing step for subsequent image processing steps. Based on these results, joint methods are developed which include the super-resolution step in the main processing task. Starting with a method for joint motion compensation and super-resolution for long-range or mobile cameras in Section 6.2, a framework for joint super-resolution and segmentation is introduced in Section 6.3.

6.1 Comparison of Super-Resolution Methods for 2D/3D Images

Instead of assuming a certain geometry when resizing depth maps, it is also possible to assume that color and depth coincide, i.e. close pixels with similar color have also similar depth. An common approach for this purpose is cross-bilateral filtering, but there are several possibilities of how to apply it and how to choose the parameters. In the following several different approaches are discussed and compared in theoretical and practical settings.

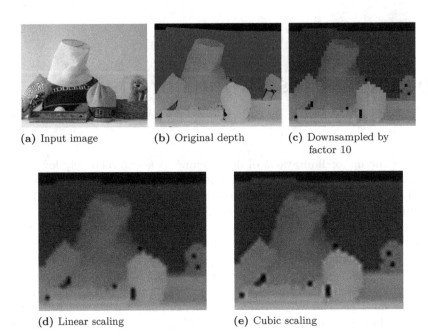

(a) Input image   (b) Original depth   (c) Downsampled by
                                                          factor 10

(d) Linear scaling   (e) Cubic scaling

Figure 6.1: Illustration of the deficiencies of standard techniques to resize
          depth images. These color and depth images are part of the Mid-
          dlebury dataset [57].

Additionally, bilateral filtering is a computationally expensive procedure and
therefore, it is also addressed how to restrict the bilateral filtering to region of
interest that accounts for the nature of the depth values gathered from a low
resolution depth imaging chip. Furthermore, recent depth super-resolution
methods based on Markov Random Fields (MRF) and a methods employing
a cost volume to judge depth assignments are discussed and evaluated.

This comparison of super-resolution methods for 2D/3D images was previously published in [L2, L4].

6.1.1 Related Work

Crabb et al. applied in [15] a bilateral filter for depth augmented alpha
matting, which is also the focus of the work of Wang, cf. [65]. A preliminary
foreground described in terms of probabilities is generated with the help of a
dividing plane in space. The closer the pixel is to the point of view the more

likely it is that belongs to the foreground. A bilateral filter is then applied on this alpha matte to fuse the depth and color information. Schuon et al. proved the ability of bilateral filtering to deal with geometric objects in [59] and in [10] a variant designed to handle noise and invalid measurements is presented.

Yang et al. define a cost function or cost volume in [68], which defines the cost of in theory all possible refinements of the depth value associated with a color pixel. Again a bilateral filter is applied on this volume and after sub-pixel refinement a proposed depth is derived. The optimization is performed iteratively to achieve the final depth map. The incorporation of a second view is also discussed. Bartczak and Koch presented a similar method using multiple views, see [1]. An approach working with multiple depth images is described in [54]. The data fusion is formulated here in a statistical manner and modeled using Markov Random Fields, on which an energy minimization method is applied. Earlier Diebel and Thrun presented a similar method in [16], which operates on a color and a depth image.

Another advanced method to combine range and color information was introduced by Lindner in [42]. It is based on edge-preserving bi-quadratic upscaling and incorporates a special treatment of invalid measurements.

6.1.2 Bilateral Filtering

Firstly, the technique of bilateral filtering will be introduced and later it is discussed how to apply it to 2D/3D images. Finally, more complex approaches are briefly explained.

6.1.2.1 Fundamentals

The method of bilateral filtering was introduced by Tomasi in [62] and it calculates for each pixel a weighted mean of all other pixels. The weights depend on the distance in space between both pixels and their photometric similarity. In practice, only neighboring pixels are involved in the calculation (meaning the measure is truncated in space). Let the vector $\underline{x} = [x, y]^T$ denote the position of a pixel and let a (vectorial) signal $\underline{s}(\underline{x}) = [s_1(\underline{x}), s_2(\underline{x}), ..., s_n(\underline{x})]^T$ represent some (vectorial) value at that position, e.g. the RGB color values and the (measured) distance. Then the weighted mean $\underline{\mu}_{\underline{s}(\underline{x})}$ for a pixel $\underline{x}$ is given by

$$\underline{\mu}_{\underline{s}(\underline{x})} = \frac{\sum\limits_{\underline{x}_i \in N(\underline{x})} \underline{s}(\underline{x}_i) \cdot h\left[\underline{g}(\underline{x}), \underline{g}(\underline{x}_i)\right]}{\sum\limits_{\underline{x}_i \in N(\underline{x})} h\left[\underline{g}(\underline{x}), \underline{g}(\underline{x}_i)\right]} \tag{6.1}$$

with $N(\underline{x})$ representing some neighborhood of the pixel $\underline{x}$ and $h[\cdot, \cdot]$ being the weighting function depending on the distance between the pixels of interest. The summation involves all pixels $\underline{x}_i$ within the specified neighborhood. $\underline{g}(\cdot)$ is some functional mapping of the pixel, depending on the pixel position and optionally also depending on the pixel value. If it depends only on the pixel position (e.g. $\underline{g}(\underline{x}) = \underline{x}$), the function is a conventional space variant or space invariant filter ($h\left[\underline{g}(\underline{x}), \underline{g}(\underline{x}_i)\right] = h[\underline{x} - \underline{x}_i]$), defined in the space domain. If $\underline{g}(\cdot)$ additionally depends on the signal value itself (e.g. $\underline{g}(\underline{x}) = \left[\underline{x}^T, \underline{s}(\underline{x})^T\right]^T$), the filter operates in the space/similarity domain. For scalar valued weighting functions a multivariate Gaussian function expressed in terms of the vector $\underline{g}$ is a good choice:

$$h\left[\underline{g}(\underline{x}), \underline{g}(\underline{x}_i)\right] = \\ \exp\left\{-\left(\underline{g}(\underline{x}) - \underline{g}(\underline{x}_i)\right)^T \cdot \Pi^{-1} \cdot \left(\underline{g}(\underline{x}) - \underline{g}(\underline{x}_i)\right)\right\} \tag{6.2}$$

with a covariance matrix Π. Instead of defining the neighborhood directly with the Euclidean distance in space

$$N_d^2(\underline{x}) = \{\underline{\hat{x}} : \|\underline{x} - \underline{\hat{x}}\|_2 < d\} \tag{6.3}$$

it can now also be specified using the weighting function $h[\cdot, \cdot]$

$$N_\varepsilon(\underline{x}) = \{\underline{\hat{x}} : h\left[\underline{g}(\underline{x}), \underline{g}(\underline{\hat{x}})\right] > \varepsilon\} \tag{6.4}$$

with a threshold $\varepsilon \leq 1$, thus involving all dimensions of the signal in the specification of the neighborhood. This corresponds to a truncated weighting function $h[\cdot, \cdot]$. Now assuming that $\underline{s}(\underline{x}) = [\underline{x}, s_1(\underline{x}), s_2(\underline{x}), ..., s_n(\underline{x})]^T = [\underline{x}, \underline{\hat{s}}(\underline{x})]^T$, $\underline{g}(\underline{x}) = \underline{s}(\underline{x})$ and $\Pi = \begin{bmatrix} I \cdot \sigma_{space} & 0 \\ 0 & \Pi_{signal} \end{bmatrix}$, where I is the identity matrix corresponding in rank to the dimension of $\underline{x}$, the multivariate weighting function separates into the product of two Gaussian kernels:

$$h\left[\underline{g}(\underline{x}), \underline{g}(\underline{x}_i)\right] = \exp\left\{-\frac{(\underline{x} - \underline{x}_i)^T(\underline{x} - \underline{x}_i)}{\sigma_{space}^2}\right\}$$

$$\cdot \exp\left\{-(\underline{\hat{s}}(\underline{x}) - \underline{\hat{s}}(\underline{x}_i))^T \cdot \Pi_{signal}^{-1} \cdot (\underline{\hat{s}}(\underline{x}) - \underline{\hat{s}}(\underline{x}_i))\right\}. \qquad (6.5)$$

Hence it is possible to specify the neighborhood radius d such that $N_\varepsilon(\underline{x}) \subset N_d^2(\underline{x})$ by $d = \sqrt{-\sigma_{space}^2 \log \varepsilon}$, which reduces the computational complexity, since there are less than $4d^2$ pixels $\underline{x}_i \in N_d^2(\underline{x})$ and these pixels can be directly addressed.

Another typical efficiency assumption is restricting Π_{signal} to be diagonal $\Pi_{signal} = diag\left(\sigma_1^2, \sigma_2^2, ..., \sigma_n^2\right)$, resulting in an independence of the signal dimensions. The different natures of space, depth measurements and color values can be accounted for by choosing different smoothing values $\sigma_{space}^2, \sigma_1^2, \sigma_2^2, ..., \sigma_n^2$ in the weighting function.

In the examples the signals $\underline{s}(\underline{x})$ are of the form $\underline{s}(\underline{x}) = [x, y, d(\underline{x}), r(\underline{x}), g(\underline{x}), b(\underline{x})]^T$ with $d(\underline{x})$ being the distance measurement and $r(\underline{x}), g(\underline{x}), b(\underline{x})$ denotes the RGB color value for a pixel $\underline{x}$. Missing values of $d(\underline{x})$ caused by a lower sampling rate are interpolated. The mapping $g(\cdot)$ is simply set to the complete signal value $\underline{g}(\underline{x}) = \underline{s}(\underline{x})$ and $\Pi = diag(\sigma_{space}^2, \sigma_{space}^2, \sigma_d^2, \sigma_{color}^2, \sigma_{color}^2, \sigma_{color}^2)$. It is usually advantageous to weight the color components differently or to use the L*a*b or the Luv color space to make differences between color values perceptually uniform.

6.1.2.2 Iterative Application

The bilateral filter is often applied iteratively to increase its effects and different approaches are possible. Firstly, only the depth values $d(\underline{x})$ can be refined though the bilateral filter. This means in practice that in each step the depth map produced in the previous step and the original image are supplied to the bilateral filter. For a pixel $\underline{x}$ and a recorded signal $\underline{s}(\underline{x}) = \underline{s}^{(0)}(\underline{x})$ this leads to the update formula for filtered signal values $\underline{s}^{(1)}(\underline{x}), \underline{s}^{(2)}(\underline{x}), ...$

$$\underline{s}^{(j+1)}(\underline{x}) = \underline{s}^{(0)}(\underline{x}) \cdot diag(1, 1, 0, 1, 1, 1)$$
$$+ \underline{\mu}_{\underline{s}^{(j)}}(\underline{x}) \cdot diag(0, 0, 1, 0, 0, 0). \qquad (6.6)$$

It should be mentioned that the mapping $\underline{g}(\underline{x})$ in $\underline{\mu}_{\underline{s}^{(j)}}(\underline{x})$ uses the updated values of $\underline{s}^{(j)}(\underline{x})$, i.e. $\underline{g}(\underline{x}) = \underline{g}^{(j)}(\underline{x}) = \underline{s}^{(j)}(\underline{x})$.

Another possibility is to refine all measured dimensions. This can be written with the recursion formula

$$\underline{s}^{(j+1)}(\underline{x}) = \underline{s}^{(0)}(\underline{x}) \cdot diag\,(1,1,0,0,0,0)$$
$$+ \underline{\mu}_{\underline{s}^{(j)}}(\underline{x}) \cdot diag\,(0,0,1,1,1,1)\,. \qquad (6.7)$$

Finally, the filter can operate on all dimensions simultaneously leading to

$$\underline{s}^{(j+1)}(\underline{x}) = \underline{\mu}_{\underline{s}^{(j)}}(\underline{x})\,. \qquad (6.8)$$

When processing each point independently from all others, i.e.

$$\underline{\mu}_{\underline{s}^{(j)}}(\underline{x}) = \frac{\displaystyle\sum_{\underline{x}_i \in N(\underline{x})} \underline{s}^{(0)}(\underline{x}_i) \cdot h\left[\underline{s}^{(j)}(\underline{x}), \underline{s}^{(0)}(\underline{x}_i)\right]}{\displaystyle\sum_{\underline{x}_i \in N(\underline{x})} h\left[\underline{s}^{(j)}(\underline{x}), \underline{s}^{(0)}(\underline{x}_i)\right]}\,, \qquad (6.9)$$

this procedure is equivalent to the Mean-Shift algorithm (see [14]).

The coordinates of the signal are not changed in the first two approaches (Eq. 6.6 and 6.7) and hence it is possible to address the neighboring pixels directly. The complexity is then $O(d^2)$ for each iteration with d being the radius of the neighborhood $N_d^2(\underline{x})$. In the last case (Eq. 6.8) the neighborhood is unknown if the pixels are processed simultaneously. Therefore, the signals for all pixels have to be checked in a trivial implementation, although space partitioning techniques such as kd-tree (cf. [3]) or a spatial registration of the filtered signals using dynamic arrays can be applied to reduce the complexity.

6.1.2.3 Cost Volume Optimization Problem

Yang in [68] and Bartczak in [1] assign a quadratic and truncated cost to each (discrete) depth change of a pixel. Let $d_1, \ldots, d_m$ be all possible depth values $d(\underline{x})$ for a pixel $\underline{x}$. Then the cost function $c\,(\underline{x}, d_k)$ which assigns a cost to a change of the depth $d\,(\underline{x}) = d_k$ is given by

$$c\,(\underline{x}, d_k) = \min\left\{\gamma, (d(\underline{x}) - d_k)^2\right\}\,, \qquad (6.10)$$

where γ is a truncating threshold. This cost function spans the initial cost volume $C^{(0)} = \left(c\left([x,y]^T, d_k\right)\right)_{xyd_k}$. The costs $C^{(\cdot)}_{\cdot\cdot d_k}$ to refine all pixels to a certain depth d_k are called a slice and a bilateral filter is applied iteratively on each slice. In the j-th iteration a signal for the slice $C_{..d_k}^{(j)}$ is given by

$$\underline{s}^{(j)}(\underline{x}) = \left[x, y, d(\underline{x}), r(\underline{x}), g(\underline{x}), b(\underline{x}), C_{\underline{x}d_k}^{(j)} \right]^T \qquad (6.11)$$

and the mapping is again $\underline{g}(\underline{x}) = \underline{s}(\underline{x})$. The smoothing values in Eq. 6.1 are $\Pi = diag\left(\sigma_{space}^2, \sigma_{space}^2, 0, \sigma_{color}^2, \sigma_{color}^2, \sigma_{color}^2, 0\right)$. The filtered cost volume in iteration $(j+1)$ is determined by

$$C_{\underline{x}d_k}^{(j+1)} = \underline{\mu}_{\underline{s}^{(j)}}(\underline{x}) \cdot [0, 0, 0, 0, 0, 0, 1]. \qquad (6.12)$$

Now the filtered signal can be calculated with

$$\underline{s}^{(j+1)}(\underline{x}) = \left[x, y, \widehat{d}(\underline{x}), r(\underline{x}), g(\underline{x}), b(\underline{x}), C_{\underline{x}d_k}^{(j+1)} \right]^T \qquad (6.13)$$

$$\widehat{d}(\underline{x}) = \arg\min_{d_k} \left\{ C_{\underline{x}d_k}^{(j+1)} \right\}. \qquad (6.14)$$

Alternatively, the local minimum of a quadratic function that is fitted to the minimum $\widehat{d}(\underline{x})$ and its neighbors can be used in the refined signal to achieve a sub-pixel refinement.

The cost volume method is based on the assumption that the scene is piecewise planar and therefore the cost of a pixel for a certain depth is correlated to the cost of refining the neighboring pixels to the same depth.

The minimum cost for a pixel can be the starting point in the next iteration used as current depth estimate.

6.1.2.4 Markov Random Field Optimization

Another possibility to construct a high resolution depth map is to formulate this task in a Markov Random Field (MRF) and to maximize the posterior probability as is described by Diebel and Thrun in [16]. This posterior probability depends on the squared difference between the generated depth map and the available depth measurements as well as on the squared difference between the depth values of neighboring pixels weighted by their photometric similarity. This methods is again based on the assumption that photometric and depth similarities coincide.

The logarithmic posterior is minimized with the conjugate gradient method. In each optimization step the photometric weights have to be determined multiple times (in the line search). Therefore, each iteration corresponds to multiple iterations in the bilateral filtering methods described above.

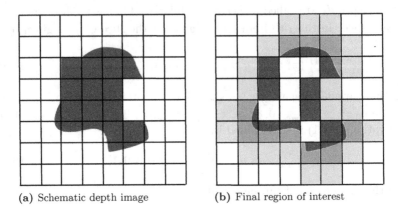

(a) Schematic depth image   (b) Final region of interest

Figure 6.2: Schematic view of a foreground object covering several depth pixel
blocks and the resulting region of interest, to which the bilateral
filter shall be applied.

6.1.3 Region of Interest for 2D/3D Imaging

As mentioned before, bilateral filtering is a computationally expensive procedure. However, in many image processing tasks only certain parts of an
image are actually of interest and form the so-called region of interest (ROI).
Applying a bilateral filter only to the ROI must be performed cautiously.
Obviously, iterative calculations directly depend on the results of all neighboring pixels. However, in theory the information of each pixel propagates
throughout the whole image after a certain number of iterations. In practice,
the influence of non-neighboring pixels is negligible for typical values of
σ_{space}. Therefore, in addition to the pixels of the ROI at least all their
neighbors should be processed as well.

When operating with depth maps with a much lower resolution than the
color images, it is essential to take the characteristics of the low resolution
modality into account when determining which area of the image should be
filtered in addition to the ROI.

Two different cases can be considered: Either a high resolution ROI is
available which was constructed at least partly based on the color image or a
low resolution ROI was determined based on the depth data. In both cases
the ROI should be enlarged by all pixels that share the same depth pixel as
a pixel in the ROI. This can be interpreted as filling the low resolution depth
pixels. Additionally, all neighboring depth pixels should be included, because

when a depth pixel covers a border of an object it is unknown whether it belongs to the background or the object. If iterative filtering is applied, all pixels in the neighborhood described with the neighborhood radius are to be processed as well. If only the border of an object is to be refined, inner pixel blocks can be neglected, see Fig. 6.2 for an illustration. When considering the computational cost of bilateral filtering, possible implementations on a GPU should be mentioned as well and will be discussed further in the next section.

6.1.4 Experiments

The different methods described in Section 6.1.2 to scale the low resolution depth images recorded with a depth camera are compared qualitatively and quantitatively in this section. Since the ground truth, i.e. a high resolution depth image, is normally not available or can only be derived by hand, the Middlebury dataset (cf. [57]) was utilized in the experiments in addition to real camera images with handmade ground truth. The following methods are evaluated: Depth refinement (Eq. 6.6), depth and color refinement (Eq. 6.7), pixelwise (Eq. 6.8, 6.9), cost volume (Eq. 6.11 and 6.12) and MRF optimization.

In Fig. 6.8 the iterative development of the refined depth images for an image of the Middlebury dataset using the different methods with common parameters is displayed. The available ground truth depth image was downsampled by a factor of 10 to gain a low resolution depth image, which serves as the starting point for the super-resolution methods. Here $\sigma_{space} = 10$, $\sigma_{color} = 30$, $\sigma_d^{-1} = 0$ and 10 iterations were used. The resulting depth maps are then compared to the original depth image. The results for different images are given in Fig. 6.3. Values for the downsampled depth image and linearly as well as cubically scaled images are plotted for comparison. A wide variety of parameters were applied to several images from the dataset to ensure that the presented results are representative. In Fig. 6.4 the iterative development for the different methods is illustrated and in Fig. 6.5 different downscaling factors are compared.

Thereafter, the methods described are tested on images recorded with the already mentioned MultiCam. An example is displayed in Fig. 6.6 with a radius of 10 and again, the iterative development is shown in Fig. 6.9. A ground truth was created for a few frames per hand and it was used to evaluate the filtered depth images. The differences (SSD) for different radii are also summarized in Fig. 6.3 as well as the results for simple upscaling techniques.

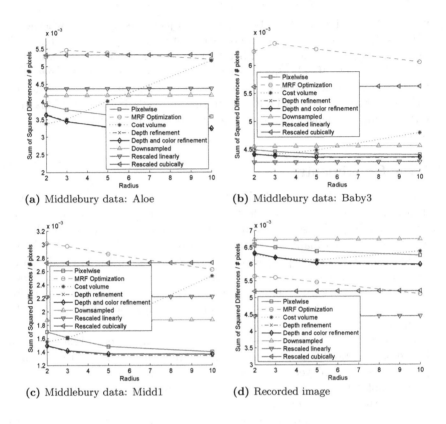

(a) Middlebury data: Aloe   (b) Middlebury data: Baby3

(c) Middlebury data: Midd1   (d) Recorded image

Figure 6.3: Comparison of the generated depth images with the original depth or the handmade ground truth respectively using different radii for $\sigma_s = 10$, $\sigma_c = 30$ and a downsampling factor of 10.

The results show clearly that all methods produce depth images which are much more accurate than a simple scaling using standard techniques - the MRF method being an exception. Linear and cubic scaling blur the depth images which results in greater differences to the original depth map. The methods of depth only and depth and color refinement perform very similarly. The pixelwise approach is slightly weaker. The cost volume method produces the nicest results when judged subjectively, but it tends to overly smooth contours, in particular for higher radii. This behavior is especially problematic for recorded real images. The MRF optimization

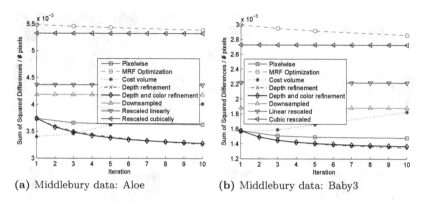

Figure 6.4: Results in different iteration steps for the discussed methods and $\sigma_s = 10$, $\sigma_c = 30$, a radius of 5 and a downsampling factor of 10.

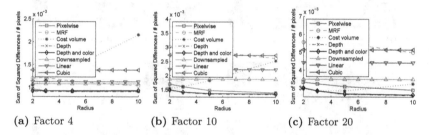

Figure 6.5: Comparison of the generated depth images with the original depth for different downsampling factors 4, 10, 20 based on the Middlebury data (Midd1) with $\sigma_s = 10$ and $\sigma_c = 30$.

method produces visibly good results, but it changes the depth values too much and thereby diverges significantly from the ground truth. The results show that $\sigma_d^{-1} = 0$, i.e. neglecting the depth values when determining the weights (cross-bilateral filtering) performs most accurately. This assertion holds true for all methods. Even in the MRF optimization ignoring the differences to the available depth measurements gives the best results. The fact that the results for the synthetic and the real data are very similar proves the validity of the results. Only the MRF optimization performs significantly better for the real scenario when using high radii, although it takes easily several minutes to process a single image in this case.

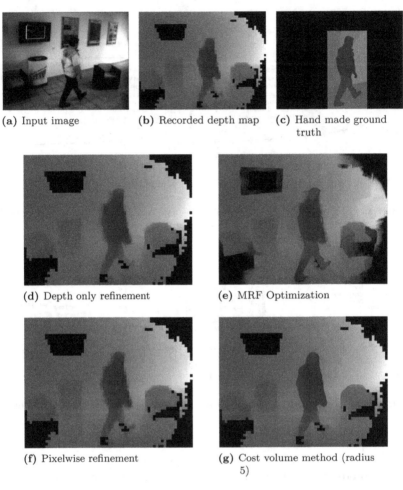

(a) Input image   (b) Recorded depth map   (c) Hand made ground
                                                          truth

(d) Depth only refinement   (e) MRF Optimization

(f) Pixelwise refinement   (g) Cost volume method (radius
                                         5)

Figure 6.6: Typical results for the discussed methods applied on the real scene
with $\sigma_s = 10$, $\sigma_c = 30$ and a downsampling factor of 10.

In Fig. 6.7 processing times per iteration are plotted for the different
methods applied to the Middlebury image (see Fig. 6.1) with a resolution
of 465×370 pixels and measured on an Intel Pentium Core2Duo processor
running at 3 GHz. Processing times for two GPU implementations of the
depth only refinement method are also displayed performed on an ATI

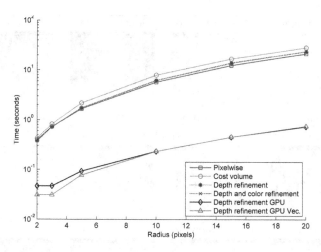

Figure 6.7: Processing time per iteration for the different filtering methods.

HD 5770 graphics card. In the first implementation all calculations for a pixel in an iteration are carried out in one thread on the GPU and the second one is a vectorized version. Here the pixelwise method performs best. The runtime of the cost volume method depends on the number of discrete depth steps (slices); here 100 depth steps were used. The processing times with only 10 slices are similar to those of the pixelwise method. As expected the GPU versions perform by more than one order of magnitude faster. These results should provide a guide for possible applications of these methods. Performance values for the MRF optimization are omitted due to the large dependence on the conjugate gradient method, for which many implementations are possible.

6.1.5 Summary

In this section different bilateral filtering techniques were formulated and compared based on laboratory tests as well as using real 2D/3D recordings. When working with 2D/3D images the different types of data usually have to be fused. In a monocular setup this can simply be achieved by rescaling using common scaling techniques, otherwise the images have to be registered first. The evaluation shows that all described bilateral filtering methods can be applied to scale the depth images and produce more accurate results than trivial scaling methods. Additionally, the cost volume-based method

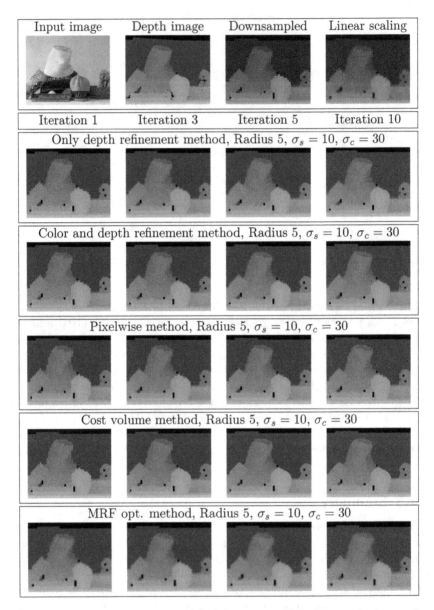

Figure 6.8: Resulting depth images after applying the super-resolution methods on a complete input image and a low resolution depth image using different parameters.

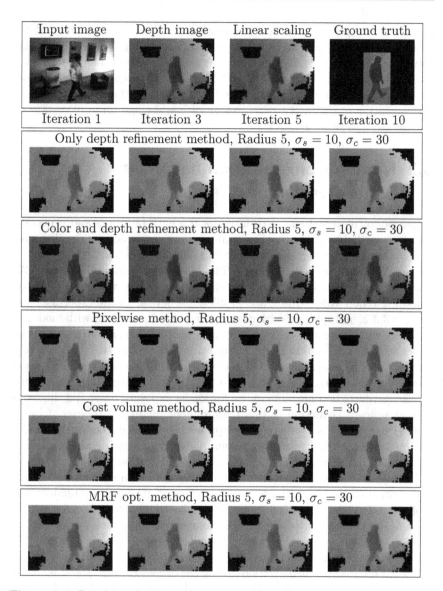

Figure 6.9: Resulting depth images after applying the super-resolution methods on a real recorded input image and a low resolution depth image using different parameters.

results in sharp but often also too smooth contours. The depth and the depth and color refinement methods perform very similarly and better than the pixelwise method. However, the latter is the fastest method and the computational demand of these methods should be taken into consideration as well. Two potential solutions to reduce processing times were discussed: the limitation of the filtering to a region of interest and an implementation on special hardware such as GPUs.

6.2 Motion Compensation and Joint Super-Resolution

When operating a camera with large focal length outdoors and especially with a large lighting device, the acquired image sequence will be distorted due to wind and other effects, since even small vibrations or movements cause image shifts of several pixels. This generally prohibits long-range 3D imaging, since long exposure times or many acquisitions with short exposure times are required to achieve acceptable noise levels. In the experiments in Section 3.3.2, multiple acquisitions under very good conditions with moderate exposure times were analyzed for long-range still imaging, since long exposure times lead to over-exposure during daytime. A similar situation is found on mobile platforms, since typical acquisition times lie between 10 ms and 50 ms and the camera can be subject to significant motion during that time. Motion artifacts become more severe for longer exposure times and this leads to a significant reduction of the measurement quality. Moreover, when a multi-frequency phase unwrapping approach is applied in order to extend the measurement range for medium and long-range imaging as proposed in Section 3.4, motion artifacts become a severe limitation.

In [44] Lottner et al. analyze motion artifacts encountered for sufficiently fast motion of objects in the scene when using PMD cameras. Lindner and Kolb introduce in [40] motion compensation based on optical flow for moving objects in videos acquired with PMD cameras. In an industry environment motion compensation is demonstrated by Hussmann et al. in [29] and [30]. These works do not consider camera movement explicitly in contrast to the proposed approach, which allows reducing the necessary computations. Axial motion of ToF cameras is discussed in [17] using the Windowed Discrete Fourier Transform. The methods for motion compensation proposed in this section were published in [L11, L13].

Camera motion during an image acquisition results in a blur for color imaging as well as motion artifacts for PMD imaging. These artifacts are invalid depth measurements and can in general attain any value. Motion artifacts are actually caused by camera motion during the acquisition of a phase image or in between the acquisition of the phase images. The former cause is neglected here due to the relatively short exposure times applied outdoors. Two features in the scene induce motion artifacts. The first one are of course edges which result in different phase shifts. Secondly, different reflectivity also introduces motion artifacts, since the differences between both PMD channels do not get canceled out then (in the nominator and denominator of the four phase algorithm, cf. Eq. 3.9). An approach to compensate camera motions in between multiple phase images is proposed in the following. The 2D/3D MultiCam allows to perform highly accurate motion estimation based on the high resolution color images instead of the low resolution PMD images. Additionally, it is possible to generate a super-resolution depth map and even to remove some of the motion artifacts resulting from shifted phase images by means of oversampling. In Section 6.2.1 a motion estimation technique is detailed, which operates in the frequency domain and produces reliable motion estimates for shifts encountered in experiments. For simplicity only lateral shifts are considered but the approach can be easily extended to handle rotations as well. Super-resolution of depth maps could be performed by a simple interpolation, but this approach suffers from the effects of the remaining motion artifacts and flying pixels. To overcome these problems, a locally weighted averaging is applied in Section 6.2.2.

The whole acquisition procedure is outlined as follows: Firstly, the camera motions are estimated based on the color images. Then the motion vectors between the frames are interpolated and mapped to each phase image. Afterwards the super-resolved image is calculated for all pixels and then the standard PMD computations to obtain depth, modulation amplitude and grayscale image of the phase images are performed. Ambiguities for long-range measurements can be resolved with the phase unwrapping approach mentioned in Section 3.4. A linear motion of the camera is assumed for the interpolation of the motion estimates.

Using individual motion estimates for phase images is valuable especially if the acquisition of the color image and the later phase images do not or only partly overlap in time, i.e. when applying a small exposure time for the color images compared to the exposure time for the PMD chip. However, the fusion of shifted phase images cannot be performed directly, since PMD chips are not intensity-calibrated. A method for intra-pixel calibration, see Section 4.2, must be applied in advance.

Another possibility is to estimate the camera motion based on the (normalized) intensity values of a phase image. This is in particular useful in order to achieve high frame rates with limited computational power and for real-time processing. Therefore, this approach is ideal for mobile platforms and will be explained in more detail in Section 6.2.4.

6.2.1 Motion Estimation

Vandewalle et al. introduced a method for planar motion estimation based on the Fourier Transformation in [63]. Since fast rotations around the roll axis are rarely observed in typical situations, the method is reduced to lateral image shifts for an increased robustness as well as to avoid aliasing when rotating images.

Let $f(\underline{x})$ be the reference image and $g(\underline{x})$ the actually observed one with $g(\underline{x}) = f(\underline{x} + \Delta \underline{x})$ with Fourier Transformations $F(\xi)$ and $G(\xi)$ for a spatial frequency $\xi \in \mathbb{C}^2$. Then the following holds

$$G(\xi) = \int g(\underline{x}) e^{-i2\pi \xi^T \cdot \underline{x}} d\underline{x} \tag{6.15}$$

$$= \int f(\underline{x} + \Delta \underline{x}) e^{-i2\pi \xi^T \cdot \underline{x}} d\underline{x} \tag{6.16}$$

$$= \int f(\hat{\underline{x}}) e^{-i2\pi \xi^T \cdot (\hat{\underline{x}} - \Delta \underline{x})} d\hat{\underline{x}} \tag{6.17}$$

$$= e^{i2\pi \xi^T \cdot \Delta \underline{x}} \cdot \int f(\hat{\underline{x}}) e^{-i2\pi \xi^T \cdot \hat{\underline{x}}} d\hat{\underline{x}} \tag{6.18}$$

$$= e^{i2\pi \xi^T \cdot \Delta \underline{x}} \cdot F(\xi) \tag{6.19}$$

with a substitution $\hat{\underline{x}} = \underline{x} + \Delta \underline{x}$.

Therefore, the phase difference $e^{i2\pi \xi^T \cdot \Delta \underline{x}}$ can be obtained given $F(\xi)$ and $G(\xi)$ with

$$\xi^T \cdot \Delta \underline{x} = \frac{1}{2\pi} \arg \left(\frac{G(\xi)}{F(\xi)} \right) . \tag{6.20}$$

Instead of computing $\Delta \underline{x}$ with for a single ξ, Vandewalle et al. argue to construct a system of linear equations for a set of frequencies and to use the least squares estimate for $\Delta \underline{x}$ in order to avoid aliasing.

6.2.2 Over-Sampling

In the following a method to combine shifted images by local averaging is presented. The images can be processed depth images, modulation amplitude and grayscale images or phase images, which are processed after motion compensation. In this case every phase $0, \pi/2, \pi, 3\pi/2$ is handled individually.

Let $\rho \in \mathcal{R}^+$ be the super-resolution factor, S the set in input images and let Δ^I be the vectorial motion estimate for the image $I \in S$. Let further $N_I(\underline{y})$ be the set of lattices of pixels in the neighborhood of lattice $\underline{y} \in I$ (with zero-based lattices). The Euclidean distance is used in the experiments to specify the neighborhood. Lastly, let $\delta_I(\underline{y})$ represent the pixel value at lattice $\underline{y}$ of image I and let h be a bandwidth parameter. Now the weighted

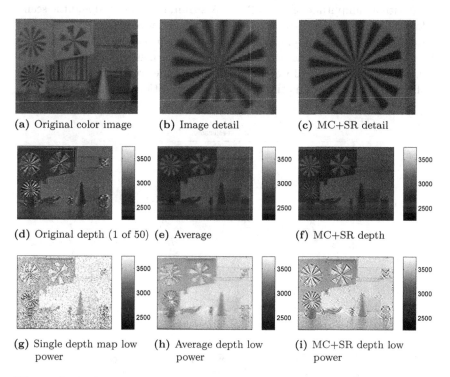

(a) Original color image   (b) Image detail   (c) MC+SR detail

(d) Original depth (1 of 50) (e) Average   (f) MC+SR depth

(g) Single depth map low power   (h) Average depth low power   (i) MC+SR depth low power

Figure 6.10: Joint motion compensation and super-resolution (MC + SR) experiment with normal lighting and with reduced lighting to simulate large distances.

local average for a pixel value $\delta_R(\underline{x})$ at lattice $\underline{x}$ in the super-resolved image R is given by

$$\delta_R(\underline{x}) = \frac{\sum_{I \in S} \sum_{\underline{y} \in N_I \left(\frac{x}{\rho} - \Delta^I - \frac{1}{2} \right)} \delta_I(\underline{y}) \cdot g(\underline{x}, \underline{y})}{\sum_{I \in S} \sum_{\underline{y} \in N_I \left(\frac{x}{\rho} - \Delta^I - \frac{1}{2} \right)} g(\underline{x}, \underline{y})} \qquad (6.21)$$

$$g(\underline{x}, \underline{y}) = \exp\left\{ -\frac{1}{h} \left\| \underline{x} - \rho \left(\underline{y} + \Delta^I + \frac{1}{2} \right) \right\|^2 \right\} . \qquad (6.22)$$

6.2.3 Experiments

In Fig. 6.10 results for artificially introduced disturbances are displayed. 50 images were acquired, while the camera was moving cyclically on a lateral plane with an amplitude of about 2 cm. A 28 mm lens was used and the scene is located in 4 meters distance. The experiment was performed with low power lighting as well to emulate long distance measurements. In Fig. 6.11 the estimated lateral camera motion is plotted. A neighborhood radius of 3 and a bandwidth of $h = 2$ were applied in the super-resolution. The results confirm that motion estimation, motion compensation and super-resolution

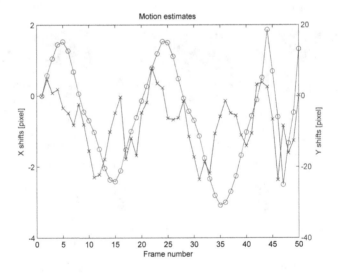

Figure 6.11: Motion estimates in X- and Y-direction for the close-range experiment.

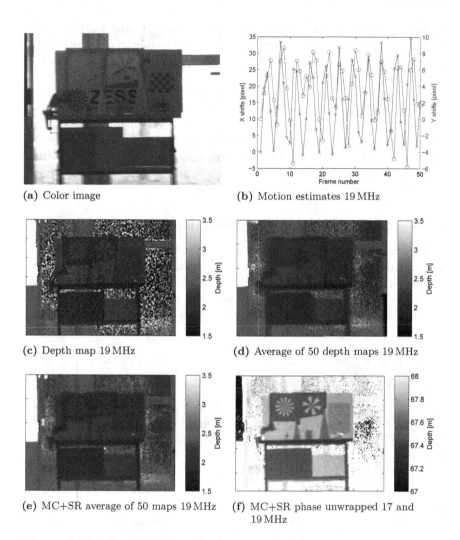

(a) Color image

(b) Motion estimates 19 MHz

(c) Depth map 19 MHz

(d) Average of 50 depth maps 19 MHz

(e) MC+SR average of 50 maps 19 MHz

(f) MC+SR phase unwrapped 17 and 19 MHz

Figure 6.12: Indoor experiment to demonstrate the motion compensation of artificial camera motions for long-range 2D/3D imaging. All depth maps were acquired with an exposure time of 1 ms and the camera motion was caused by a eccentric rotation of a mass.

can be successfully applied in both lighting conditions. Outliers at the border
of objects can be observed in the depth maps, which originate from PMD
motion artifacts and are smoothed out by the weighted averaging in the
super-resolution processing step.

A comparison of a simple averaging of depth images to the proposed joint
motion compensation and super-resolution with phase unwrapping is given
in Fig. 6.12 for a long-range test setup in 68 meters distance. Again 50
images were recorded at modulation frequencies of 17 MHz and 19 MHz with
an exposure time of 1 ms. The direct averaging produces severe blurring,
while the proposed method preserves edges and distance values and the
phase unwrapping can be included successfully.

6.2.4 Real-time Motion Compensation

High accuracy motion compensation is difficult to perform in real-time,
in particular when it serves as a pre-processing method and other image
processing tasks should be executed. In order to accomplish motion compensation (MC) in real-time, the procedure can be simplified as follows.
Let A_i and B_i with $i = 1, 2, 3, 4$ be the phase images for PMD channel

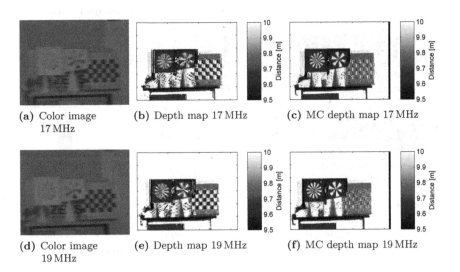

(a) Color image   (b) Depth map 17 MHz   (c) MC depth map 17 MHz
    17 MHz

(d) Color image   (e) Depth map 19 MHz   (f) MC depth map 19 MHz
    19 MHz

Figure 6.13: Results for the intra-frame real-time motion compensation involving a moving camera and two modulation frequencies.

A and B respectively acquired with phase shift i and let $I_i = A_i + B_i$ be the phase intensity image. The phase images are intensity calibrated, see Section 4.2, and the SBI effects are corrected (Section 3.1.5). With the motion estimation method detailed in Section 6.2.1 the lateral image shifts Δ_i between I_1 and I_i for $i = 2, 3, 4$ can be determined and rounded to the nearest integer. These image shift are applied to the associated phase images (can be replaced with pointer arithmeticians) and the PMD calculation are performed. Motion compensation for a single depth map working on the individual phase images is referred to as intra-frame MC as opposed to inter-frame MC, which operates on multiple depth maps.

In Fig. 6.13 an example for the intra-frame MC is given for a test setup acquired with an exposure time of 5 ms, a 50 mm lens and two different modulation frequencies. A motion between phase images of about 1.2 pixels is estimated and subsequent images with different modulation frequencies are shown. The results demonstrate that the motion artifacts can be largely

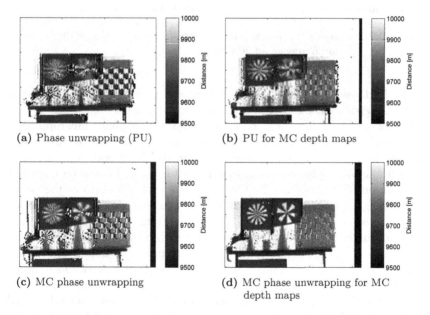

(a) Phase unwrapping (PU)

(b) PU for MC depth maps

(c) MC phase unwrapping

(d) MC phase unwrapping for MC depth maps

Figure 6.14: Demonstration of severe motion artifacts when performing phase unwrapping and results for normal and motion compensated (MC) phase unwrapping.

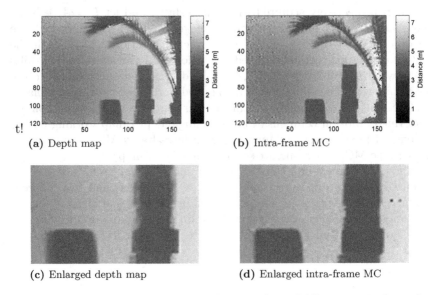

(a) Depth map   (b) Intra-frame MC

(c) Enlarged depth map   (d) Enlarged intra-frame MC

Figure 6.15: Experiment to evaluate the intra-frame MC using a single modulation frequency of 20 MHz .

reduced. However, at the edges of objects and high contrast regions some
artifacts persist, since only full pixel shift are applied.

Motion artifacts are even more prominent when phase unwrapping is
conducted. However, the motion compensated depth maps acquired with
two (or more) modulation frequencies can also be motion compensated
(inter-frame MC). The camera motion is then estimated based on both first
phase intensity images I_1. In Fig. 6.14 four cases to perform the phase
unwrapping are considered: Without motion compensation, with two motion
compensated depth maps, with two normal depth maps but with inter-frame
MC and finally with motion compensated depth maps and inter-frame MC.
The results demonstrate how important the proposed method is for medium-
range depth imaging with phase unwrapping on mobile platforms or when
other camera motions occur.

A more realistic experiment to demonstrate the intra-frame MC is displayed
in Fig. 6.15. Here a modulation frequency of 20 MHz, a 8.5 mm lens and
an exposure time of 10 ms were used. A fast motion was introduced and
results in motion artifacts at the edges of objects, which are successfully
removed when applying the MC. Only flying pixels at the edges of objects

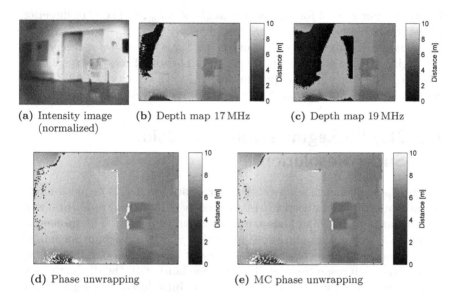

(a) Intensity image (normalized) (b) Depth map 17 MHz (c) Depth map 19 MHz

(d) Phase unwrapping (e) MC phase unwrapping

Figure 6.16: Phase unwrapping experiment on a rotation table. An image shift of 3 pixels was detected between the first phase images of both acquisitions.

remain, which cannot be corrected without affecting the contours of objects. However, a simple thresholding with the modulation amplitude usually allow to remove them.

Results for another experiment to demonstrate the inter-frame MC with a different camera, lens and illumination system are depicted in Fig. 6.16. The camera and the illumination system were mounted on a rotation table and depth maps were acquired with alternating modulation frequencies of 17 and 19 MHz and an exposure time of 5 ms. The results show how the incorrect estimation of the real distances can be reduced at discontinuities, e.g. at the doors. It should be mentioned here that we observe multi-path reflections on the reflective floor.

The inter-frame MC is not effective for slow camera motions and short acquisition times. The angular velocity v which results in an image shift of one pixel between the first and the last phase image can serve as a criterion and it is calculated with

$$v = \frac{\alpha_{cam}}{N \cdot t_{acq}} \qquad (6.23)$$

in degrees per second for an opening angle of α_{cam}, N pixels in direction of the rotation and an acquisition time of $t_{acq} = 3 * (t_{exp} + t_{ro})$. Here t_{exp} is the exposure time for a phase image and t_{ro} is the read-out time of the PMD chip. In the experiments we get for a horizontal rotation (yaw axis), the 19k PMD chip and an exposure time of 5 ms a velocity of 14.2°/s for a 8.5 mm lens and for a 50 mm lens a critical angular velocity of 2.6°/s.

6.3 2D/3D Segmentation and Joint Super-Resolution

A versatile multi-image segmentation framework for 2D/3D or multi-modal segmentation is introduced in this section with potential applications in a wide range of machine and computer vision problems. The framework performs segmentation and super-resolution jointly to account for images of unequal resolutions obtained from different imaging sensors. This allows to combine high resolution details of one modality with the distinctiveness of another modality. A set of measures is introduced to weight sensor measurements according to their expected reliability and it is employed in the segmentation as well as in the super-resolution. The approach is demonstrated with the help of different experimental setups and the effect of additional modalities as well as of varying parameters of the framework are evaluated.

6.3.1 Overview

Segmentation is a widely studied topic in the areas of image processing, computer and machine vision with many applications in associated fields. The main source of information are color images and several approaches have been proposed like pixel-based methods, edge-oriented methods, region- and texture-based approaches. Another research topic is how to apply these methods to other information sources, e.g. radar data, MRI scans or depth measurements.

The capability of a single modality to distinguish between objects is not sufficient or robust enough for some applications. The obvious solution is to utilize information obtained from an additional modality (possibly acquired with another imaging sensor). However, the measurements from the imaging sensors need to be registered, which is often non-trivial and sometimes certain assumptions are required. Additionally, it cannot assumed that the resolutions of the imaging devices match. When the lowest resolution is not sufficient for the task at hand a super-resolution method is required as a pre-processing step, since normal scaling methods are often not appropriate. In this section a segmentation and joint super-resolution framework is introduced, which utilizes a standard segmentation method (Mean-Shift algorithm) and estimates super-resolved input images in an iterative process. The proposed method incorporates validity measures to judge the measurement quality of the input data in order to account for noise and disturbances. The method is widely applicable and was published in [L6].

In Section 6.3.2 the related literature is discussed and in Section 6.3.3 relevant modalities for the experiments are described including a summary of how to device them technically. Afterwards the segmentation and super-resolution framework is detailed in Section 6.3.4. In Section 6.3.5, experimental results of this framework are reviewed and this chapter ends with a summary in Section 6.3.6.

6.3.2 Related Work

In this section research dealing with multi-modal and depth image segmentation will be reviewed. In [7] the Mean-Shift algorithm is applied to color and depth images acquired with two cameras (binocular setup). The depth images are firstly resized with a bilateral filter and then segmented. Different super-resolution methods for depth imaging including several variants of bilateral filtering are compared in [L4], see also Section 6.1. Color and depth are used for the task of alpha matting in [65].

The segmentation of depth maps is performed in [32] for compression purposes and in [33] intensity and depth information of the same size is segmented with the graph-cut method in order to detect planar surfaces. In [64] a watershed-based segmentation is utilized to analyze biological samples with 2D or 3D data and it combines intensity, edge and shape information. Lastly, the segmentation of ultrasound images in 2D or 3D is studied in [9].

6.3.3 Multi-Modal Sensor Data

In the area of machine vision standard color or grayscale images are the predominant source of information, but nevertheless, supplemental modalities play a growing role. In addition to a color chip, a Time-of-Flight imaging chip is utilized in the following, which is able to provide depth and near infrared reflectivity measurements. Remember that the lateral resolution of such imaging chips is significantly lower than those of standard color or grayscale chips and also too low to capture fine details, which are searched for in many applications. Nevertheless, these additional modalities may be able to provide valuable clues depending on the application.

The measurements often do not provide the information to be utilized in the segmentation directly. Shadows or varying lighting are effects visible to color chips, but they do not contain semantic information in many cases. Usually the same is true for depth measurements, e.g. planes parallel to the imaging plane do not have similar depth values and this will easily lead to multiple segments. Therefore, the first step is to derive the information from the images, which is useful to a segmentation task. Color images are often transformed to the L*u*v* color space to make differences perceptually uniform and to be able to eliminate the influence of lighting conditions. Regarding depth images, normal vectors of the depth image can be derived and used in the segmentation if there are mostly planar objects in the scene. In the following this transformed information will be referred to as features.

Normal vectors are estimated by firstly calculating 3D points of the depth values using a range camera model. Afterwards the surface normal $\underline{n} = (n_x, n_y, n_z)$ at each 3D point $\underline{p} = (p_x, p_y, p_z)$ can be interpolated by averaging over the 8 normal vectors of triangles spanned by $\underline{p}$ and combinations of its neighbor points. This will lead to interpolation errors at the borders of objects and may need additional processing.

Multi-modal measurements originate from different imaging devices or sources and are influenced by different imaging conditions. Therefore, these features also have a varying reliability. In some cases it is useful to utilize certain measures to judge the validity of measurements. The quality of lighting greatly influences color information and it can be estimated based on the Luminance. For amplitude modulation CW ToF imaging the modulation amplitude is a measure describing the amount of active lighting at any given point and serves as a valuable descriptor. The variance over time is for all measurements a reliable validity measure as long as it is available, see Section 6.3.4.1 for more details.

Additionally, the different images or multi-modal information need to be registered at first in order to obtain spatially corresponding measurements. With the monocular MultiCam this simply consists of applying a scale factor and an offset. When using multiple cameras, this is in general more complicated, but for some machine vision applications an affine transformation may suffice.

6.3.4 Multi-Image Segmentation Framework

Many traditional segmentation methods perform a clustering of points in the so-called feature space consisting of coordinates and measurements. The main advantage of these methods is that they are typically fast. On the downside these methods require input images of the same size. In Section 6.3.4.1 a widely applied feature space segmentation method, the Mean-Shift algorithm, is reviewed.

Furthermore, a segmentation can be utilized to generate high resolution images from the input images of lower resolution. In Section 6.3.4.2 such a super-resolution method is discussed. This leads to an estimation problem, which can be expressed in the Expectation Maximization (EM) framework as follows: all labels of feature points make up the parameter set Ω and the missing measurements due to lower resolutions are unobserved latent variables Z. This is based on the assumption that the segmentation algorithm maximizes the likelihood given Z.

1. **Initialization**: Generate uniform segmentation $\Omega^{(0)}$.

2. **Estimation**: Perform super-resolution to estimate Z.

3. **Maximization**: Perform a multi-image segmentation to retrieve $\Omega^{(t+1)}$.

The E- and M-steps are iterated until convergence is reached or a maximum number of iterations were performed. In the following sections the segmentation and super-resolution methods are detailed.

6.3.4.1 Weighted Multi-Modal Mean-Shift

The Mean-Shift algorithm [14] was explained in detail and applied to normal vectors generated of a depth map in Section 4.4.1. Here a more general weighting function is applied and it will be defined in the following. Let $P_i = \left(\underline{p}_i, \underline{f}_i \right)$ and $P_j = \left(\underline{p}_j, \underline{f}_j \right)$ be points of the feature space with lattices $\underline{p} \in \mathbb{R}^2 = \Psi$ and features $\underline{f} = \Phi$. For the weight $g\left(P_i, P_j\right)$ between two points the product of two separate Gaussian kernels is used (simpler and faster kernels are of course possible)

$$g\left(P_i, P_j\right) = \exp\left\{ -\frac{\left\|\underline{p}_i - \underline{p}_j\right\|_2^2}{h_{space}} \right\} \cdot \exp\left\{ -\frac{\left\|\underline{f}_i - \underline{f}_j\right\|_r^2}{h_{range}} \right\} . \qquad (6.24)$$

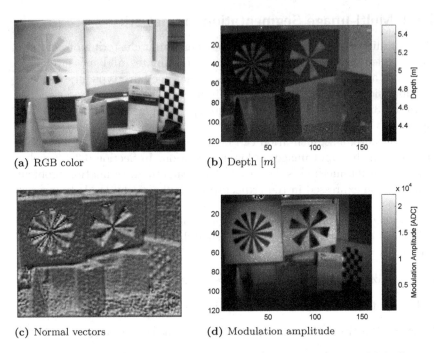

(a) RGB color (b) Depth $[m]$

(c) Normal vectors (d) Modulation amplitude

Figure 6.17: Different modalities (colorization) acquired with the MultiCam.

The first term is based on the squared Euclidean distance between the two points and the second uses a (pseudo-)norm to measure differences between features. h_{space} and h_{range} are bandwidth parameters to control the influence of the kernels.

The update formula of the Mean-Shift section can easily be extended with additional weights $\omega(P)$ leading to

$$P_i^{(t+1)} = \frac{\sum_{P \in N\left(P_i^{(t)}\right)} P \cdot \omega(P) \cdot g\left(P_i^{(t)}, P\right)}{\sum_{P \in N\left(P_i^{(t)}\right)} \omega(P) \cdot g\left(P_i^{(t)}, P\right)} . \tag{6.25}$$

The weights are specific to a feature point P and not specific to a subspaces of the feature space, since the segmentation is performed jointly over all modalities. In Fig. 6.17 different modalities acquired with the MultiCam are displayed. Let $P = (p, \underline{f})$ be a feature point and $\underline{f} = (f_L, f_u, f_v, f_d, f_{mod}, n_x, n_y, n_z, f_\sigma) \in \Phi$ with a color value (f_L, f_u, f_v) in the

L*u*v* color space, a depth value f_d, a modulation amplitude f_{mod}, a normal vector $\underline{n} = (n_x, n_y, n_z)$ and a variance f_σ of the depth measurements if multiple acquisition are available. The validity measures are defined as follows: $\gamma_{lum}(P)$ gives penalties for low and high luminance levels with an appropriate parameter, e.g. $\alpha_{lum} = 70$, and a bandwidth $h_{lum} = 2$

$$\gamma_{lum}(P) = \exp\left\{-\frac{(f_L - \alpha_{lum})^2}{h_{lum}}\right\} . \tag{6.26}$$

Similarly, the modulation amplitude of CW ToF imaging reliably describes the noise level of the depth measurements, hence another measure can be defined as $\gamma_{mod}(P) = 1 - \exp\left\{-\frac{f_{mod}^2}{h_{mod}}\right\}$ and its influence is controlled with a bandwidth parameter h_{mod}. Another indicator of the quality of the depth measurement is the variance over time, which is exploited in $\gamma_{var}(P) = \exp\left\{-\frac{f_\sigma^2}{h_\sigma}\right\}$. The complete validity weight is then given by $\omega(P) = \gamma_{lum}(P) \cdot \gamma_{mod}(P) \cdot \gamma_{var}(P)$.

In the weight $g(P_i, P_j)$ between two feature points each subspace is typically treated independently, i.e. the range (pseudo)-norm $\|\cdot\|_r$ consists of separate norms for each subspace. Euclidean norms are commonly used, but this is not appropriate for some modalities, especially when comparing two normal vectors. Here the squared sine of the enclosed angle is much more suitable. Nevertheless, for simplicity it is also possible to rely on the distance measure $\delta_n\left(\underline{n}_i, \underline{n}_j\right)$ between normal vectors $\underline{n}_i$ and $\underline{n}_j$ of unit length, which embeds an Euclidean distance in a Gaussian kernel

$$\delta_n(\underline{n}_i, \underline{n}_j) = \exp\left\{-\frac{\left(\|\underline{n}_i - \underline{n}_j\|_2^2 - 2\right)^2}{h_{normal}}\right\} . \tag{6.27}$$

In the experiments the following weighting kernel is applied to measure differences between the feature points $P_i = (\underline{p}_i, \underline{c}_i, d_i, \underline{n}_i)$ and $P_j = (\underline{p}_j, \underline{c}_j, d_j, \underline{n}_j)$ with lattices $\underline{p}_{\cdot}$, color values $\underline{c}_{\cdot}$, depth values $d_{\cdot}$ and normal vectors $\underline{n}_{\cdot}$ with associated bandwidths

$$g(P_i, P_j) = \exp\left\{-\frac{\|\underline{p}_i - \underline{p}_j\|_2^2}{h_{space}} - \frac{\|\underline{c}_i - \underline{c}_j\|_2^2}{h_{col}} - \frac{(d_i - d_j)^2}{h_{depth}}\right\} \delta_n(\underline{n}_i, \underline{n}_j) . \tag{6.28}$$

6.3.4.2 Joint Super-Resolution

A given segmentation can serve as a basis to estimate high resolution images from multi-modal images of lower resolution. Super-resolution for multiple images is often performed under the assumption that data in the different images coincides, e.g. color and depth. Cross bilateral filtering, see Section 6.1.2, works under this assumption and it is used in many approaches. However, the same notion is true for segments found in multi-modal data.

Let $S_1, S_2, \ldots, S_m \subset \Psi \times \underline{\Phi}$ be segments with $S_i \cap S_j = \emptyset$ for $i \neq j$ with a subspace $\underline{\Phi}$ of Φ, for which a super-resolution should be performed. Let $\underline{q}^{(t)} = (\underline{p}, \underline{f}^{(t)}) \in S_i$ be a point at iteration t of the EM algorithm with arbitrary lattice $\underline{p}$ and feature $\underline{f}^{(t)}$. Let further $g(\cdot, \cdot)$ be a (Gaussian) kernel and $N(\cdot)$ a spatial neighborhood. In the subsequent iteration $\underline{q}^{(t+1)}$ is computed with

$$\underline{q}^{(t+1)} = \frac{\sum_{\underline{u} \in S_i \cap N\left(\underline{q}^{(t)}\right)} \underline{u} \cdot \omega\left(\underline{u}\right) \cdot g\left(\underline{q}^{(t)}, \underline{u}\right)}{\sum_{\underline{u} \in S_i \cap N\left(\underline{q}^{(t)}\right)} \omega\left(\underline{u}\right) \cdot g\left(\underline{q}^{(t)}, \underline{u}\right)} \ . \tag{6.29}$$

There are different possibilities to calculate this formula, e.g. the spatial neighborhood can include only the actual measurements or every feature point on a finer lattice and there are many ways to choose the kernel. In the experiments with the MultiCam, feature points on the lattice of the color image are generated at first, since it has the highest resolution. The other images are transformed, which consists here only of a nearest neighbor scaling and a translation. Then the formula is applied with a spatial kernel for each feature point yielding a new set of points.

6.3.5 Experimental Evaluation

In order to evaluate the proposed segmentation and super-resolution framework, results are discussed for the well known high quality Middlebury benchmark dataset (cf. [57]), which consists of sets of color images taken from different views and associated disparity maps. One specific view of one test scene was chosen and it is depicted in Fig. 6.18, where the disparity map was converted to a depth map and the normal vectors were computed. The depth map was downsampled with a factor of 5 to simulate measurements of different lateral resolutions. One exemplary segmentation is shown and the super-resolution results are also displayed, which exhibit weaknesses for thin objects, but in general work as expected. Furthermore, the iterative estimates of the high resolution depth map are shown in Fig. 6.19. One

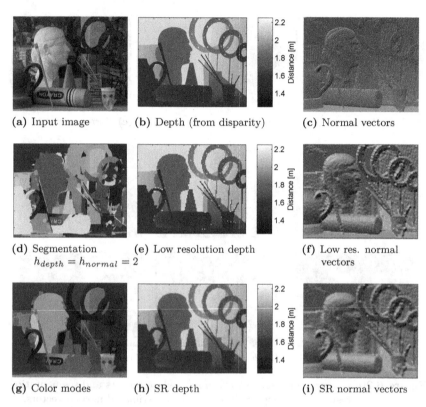

(a) Input image   (b) Depth (from disparity)   (c) Normal vectors

(d) Segmentation   (e) Low resolution depth   (f) Low res. normal
$h_{depth} = h_{normal} = 2$ vectors

(g) Color modes   (h) SR depth   (i) SR normal vectors

Figure 6.18: Multi-modal segmentation results for a benchmark image of the Middlebury dataset and super-resolution images.

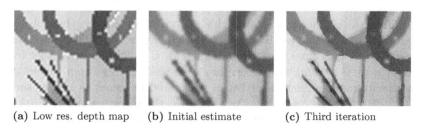

(a) Low res. depth map   (b) Initial estimate   (c) Third iteration

Figure 6.19: Incremental estimation of the super-resolved depth map.

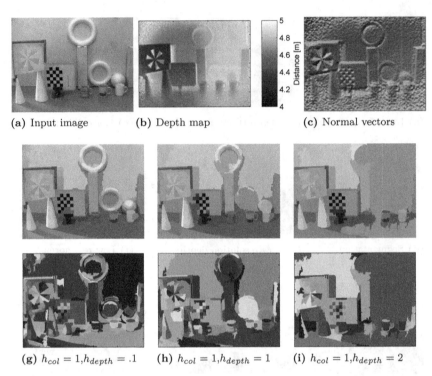

(a) Input image   (b) Depth map   (c) Normal vectors

(g) $h_{col} = 1, h_{depth} = .1$ (h) $h_{col} = 1, h_{depth} = 1$ (i) $h_{col} = 1, h_{depth} = 2$

Figure 6.20: Experimental setup with mostly white objects and segmentation results using the depth map and the estimated normal vectors.

can observe a very smooth initial estimate and a sharpening in following iterations.

In Fig. 6.20 a similar setup, in which a variety of mostly white objects are arranged, is evaluated. The setup was acquired with the MultiCam and depth as well as normal maps were calculated. It should be noted that it is usually possible to find parameters for the Mean-Shift algorithm to perform a segmentation of sufficient quality of such uniform scenes as long as some color differences are provided. Nevertheless, these segmentations are not robust and thus not reliable, since small changes to the parameters will lead to very different outcomes. The normal vectors do not provide valuable segmentation hints for this scene due to the parallel planes and curvatures. Segmentation results based on color and depth for three different sets of parameters are given with colored labels and average color of the segments.

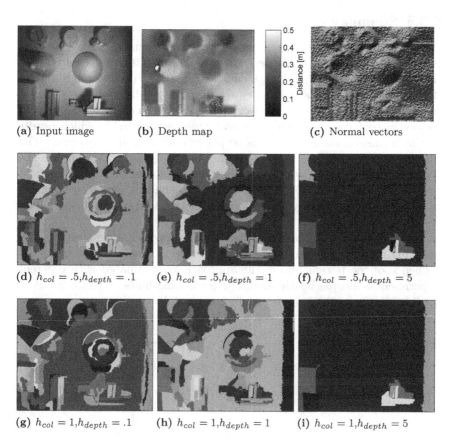

Figure 6.21: Test scene acquired under difficult lighting conditions and a set of different segmentations.

The depth bandwidth h_{depth} was changed to demonstrate the influence of the depth measurements on the results and its significant impact can be observed. In Fig. 6.21 a similar setup was acquired and it is demonstrated that in this case color information provides only small hints for the segmentation. The depth measurements can be utilized in conjunction with normal vectors to perform the segmentation and the color information is exploited mainly in the super-resolution task.

6.3.6 Summary

A modular multi-image segmentation framework for multi-modal data was introduced in this section. Since multi-modal data is usually acquired with different imaging chips, the segmentation approach needs to account for different resolutions and necessary transformations to align the different modalities. The proposed framework incorporates an estimation approach to jointly perform a segmentation and super-resolution, in which results of the segmentation influence the super-resolution and vice versa. The high resolution images generated are not only required to accomplish the segmentation at borders of objects, but can also be utilized in subsequent processing steps. A set of validity measures is defined to give measurements of expected lower quality lower weights in the segmentation and super-resolution.

The multi-modal segmentation framework proposed is demonstrated by applying it to 2D/3D segmentation. The validity measures as well as the influence of the parameters of the approach are evaluated with different experimental setups, which should give valuable hints in which areas of application the method can be applied successfully.

7 Multiple Camera 2D/3D Tracking

In this chapter a tracking approach designed to operate with multiple cameras with optional depth information, e.g. ToF cameras, structured light cameras and stereo or multi-camera setups, is introduced. The approach combines photometric tracking with volumetric tracking and it is capable of working with any number and type of cameras. In order to achieve this objective the tracked object is modeled in 3D with an ellipsoid. The density of the observed space is modeled with a set of Gaussian kernels for each line of sight in order to make use of the depth information. A proposed target configuration is then evaluated by projecting each observed color image onto the ellipsoid and comparing this projection to the expected appearance. Additionally, the density of the space occupied by the ellipsoid is estimated and compared to the expected density. It will be demonstrated that by employing the depth information in this way, ambiguities due to color similarities and changing appearance can be overcome reliably.

7.1 Overview

Tracking is one of the most important and widely spread video processing steps, which has been applied since the early days of digital video processing in the fields of motion capture, activity recognition and analysis. Applications range from security and safety systems, assisted living up to gesture recognition and other interaction methods.

The tracking algorithms of choice are the Kalman filter and the particle filter or CONDENSATION algorithm (cf. [31]), which form a widely accepted standard and define mechanisms to locate a given target based on the current observations, i.e. assigning probabilities to proposed target states.

In addition to these mechanisms tracking approaches differ in the way how they describe the target, how they measure the probability of a certain object state, how they adapt the target model over time and so on. These definitions have to be adapted to the measurement system, i.e. one or multiple color cameras, Time-Of-Flight cameras, radar and other imaging devices.

In this chapter a multi-view and multi-modal tracking approach is presented, which is able to utilize any number of color as well as 3D cameras. It is based on the particle filter framework and probabilities are assigned to object states through photometric as well as volumetric measures. A state is modeled with an ellipsoid in 3D space, which can be generated from a set of 3D points with the help of the Maximum Likelihood estimator. The density of the space on the other hand is modeled by a set of Gaussian kernels for each pixel or line of sight. The volumetric measure is defined as the difference between the expected intersection density and the actual density. The photometric measure compares the expected appearance and the observed appearance, which is calculated by projecting the input images onto the ellipsoid.

The photometric measure is evaluated for every color camera, whereas the volumetric measure is calculated for each ToF camera in the setup. All models and measures will be described in detail and some experiments will be discussed to evaluate the capabilities of the approach presented in this thesis. This approach was previously published in [L5].

In Section 7.2 the related literature is discussed. Afterwards the proposed tracking approach is presented in Section 7.3. Some experiments are discussed in Section 7.4 and the chapter closes with a conclusion in Section 7.5.

7.2 Related Work

Ghobadi et al. used the CONDENSATION algorithm in [24] to track a robot arm and a human operator in an industrial environment by clustering of depth values gained from a 2D/3D camera and in [23] this approach is compared to other techniques.

In [35] depth histograms of persons are also tracked with the help of the CONDENSATION algorithm, whereas in [7] objects are tracked based on color and depth histograms using the Mean-Shift algorithm. A method based on separate color and depth histograms working with the CONDENSATION algorithm was presented in [56]. A different approach that involves tracking was presented in [26]. Here the pose of a human arm was estimated based on a combination of the silhouette and the 3D position of the arm.

In [25] depth measurements from a ToF camera are clustered with the k-means algorithm and these clusters are compared to a training set of clusters representing head and shoulders. The actual tracking consists of correlation matching, which is also used in [5] but here blobs, i.e. connected sets of pixels, instead of clusters are used. In [67] depth and intensity images from a ToF camera are used at first to construct a background model. Then the 3D points not belonging to the background are projected onto the ground plane and clustered. These clusters are then tracked over time.

A multi-view tracking approach was presented by Khan and Shah in [36]. They deal with occlusions and ambiguities by combining the information from different cameras and prevent intersections of objects or people through a global homography constraint. Additionally, in [4] a multi-view method based on blob-tracking was demonstrated in a smart room environment. The targets were characterized by color histograms and Haar-like features.

The approaches presented in [35] and [7] are the most similar ones to the method introduced in this chapter. The main difference is that these approaches treat the distance measurement of a pixel like an additional color channel, whereas in this thesis the object and the space are explicitly modeled in 3D. Thus, it is possible to combine multiple cameras of different types globally, i.e. a single probability density function describing possible object states can be estimated and not several which have to be fused later on.

7.3 Proposed Tracking Approach

The tracking method described in this chapter is based on particle filters. A particle filter is a technique to estimate a non-Gaussian distribution over a state space. A state X defines a configuration of the tracked object, e.g. the position, size and orientation, and the probability distribution density $p_X(\xi)$ describes the probability of the realization ξ being the true configuration. This distribution is approximated with a set of weighted samples $\{s_i, \omega_i\}$ for $i = 1, \ldots, n$. The weight of a sample is determined by measuring how accurately a certain configuration s_i resembles the current observation, i.e. by evaluating the likelihood. This procedure is formalized in the CONDENSATION algorithm, which is described in detail in [31].

In Fig. 7.1 an overview of the approach with n color and m ToF cameras is illustrated. The likelihood (weight) of a certain configuration s. is calculated by projecting each color image onto the 3D-object model (here an ellipsoid) and comparing it to the appearance model of the tracking target. The appearance model consists of a photometric model (a histogram of the intersection of the projected image and the ellipsoid) and a density. In order to determine this density, the depth information obtained from the

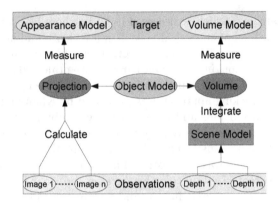

Figure 7.1: Schematic overview of the proposed tracking framework.

depth cameras is used to build up a space model, which consists of a set of Gaussian kernels for each line of sight. The intersection density of an object configuration and the 3D space can then be calculated and compared to the expected density.

In the remainder of this section at first the object and space models are defined and then the appearance models and the measuring process are described.

7.3.1 Object Model

As already mentioned, the tracked object is modeled with an ellipsoid in the proposed approach. The reason for this is that a Gaussian distribution $\mathcal{N}(\mu, \Sigma)$ with $\mu \in \mathbb{R}^3$, $\Sigma \in \mathbb{R}^{3 \times 3}$ can be easily created from set of 3D points by employing the Maximum-Likelihood estimator. When defining a fixed standard deviation or Mahalanobis distance d, this Gaussian describes an ellipsoid.

In the following a distance measure $\varphi(\cdot, \cdot)$ between two samples or states will be detailed. The distance between two states described by $\mathcal{N}(\mu_1, \Sigma_1)$ with a volume of V_1 and $\mathcal{N}(\mu_2, \Sigma_2)$ with a volume of V_2 is measured through the Euclidean distance between their centers, their relative difference in volume and with an intersection volume estimate $I(\mathcal{N}(\mu_1, \Sigma_1), \mathcal{N}(\mu_2, \Sigma_2))$ in relation to $\max\{V_1, V_2\}$. The intersection volume $I(\cdot, \cdot)$ is estimated by randomly generating N points on the hull of the smaller ellipsoid and checking how many of these are inside the larger ellipsoid and, of course, this measure has to be truncated. The similarity measure is then given by

$$
\varphi((\mu_1, V_1),(\mu_2, V_2)) = \exp\left(-\frac{\|\mu_1 - \mu_2\|^2}{2\sigma_{space}^2}\right)
$$
$$
\cdot \exp\left(-\frac{1}{2\sigma_{size}^2}\log\left(\frac{\min\{V_1, V_2\}}{\max\{V_1, V_2\}}\right)^2\right)
$$
$$
\cdot \exp\left(-\frac{1}{2\sigma_{is}^2}\log\left(\frac{I(\mathcal{N}(\mu_1, \Sigma_1), \mathcal{N}(\mu_2, \Sigma_2))}{\max\{V_1, V_2\}}\right)^2\right) \tag{7.1}
$$

with tuning constants σ_{space}, σ_{size} and σ_{is}.

For simplicity the so-called stochastic diffusion, which is applied in order to generate new samples, is only outlined. The two distributions describing the differences in volume and space are randomly sampled and the ellipsoid is moved and resized accordingly. Afterwards the ellipsoid is rotated by sampling randomly from a Gaussian distribution depending on another tuning constant σ_{angle}.

7.3.2 Modeling the Density of 3D Space

The density of space is modeled in 3D with the help of depth information. In the current implementation, depth measurements are performed with ToF cameras, which makes it possible to model the observed density for each line of sight. Each 3D point on a line of sight is assigned a certain probability that it contains mass. This probability is estimated based on the information if a point in space is reflecting light. If this is the case, it is assumed that an object is located there, see Fig. 7.2 for plots of the space model for a real video sequence. The space model describes the density of each line of sight with a set of Gaussian kernels. The idea is to represent current and previous observations simultaneously in order to represent the foreground and background at the same time.

The variances of the Gaussians provide the means to handle different levels of noise, e.g. in areas where the depth measurements are extremely noisy the variance will be high and hence the density of such a line of sight will be assumed to be also high. This will reduce the influence of the volumetric measure, since different depth values will not change the measured density significantly in this case. For the space behind the Gaussian with highest depth value, a minimum density probability of $\gamma_{unknown}$ is assumed.

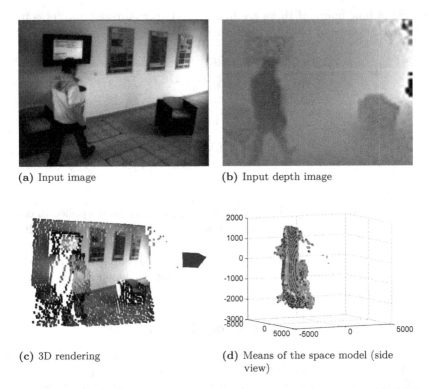

(a) Input image   (b) Input depth image

(c) 3D rendering   (d) Means of the space model (side
                                        view)

Figure 7.2: 3D model of space built from a recorded video using a single Multi-
Cam.

The probability of mass existing at depth d for a certain line of sight is
estimated with

$$p_{xy}(d) = \max_i \{V(d - \mu_i)\} \tag{7.2}$$

$$V(\tilde{d}) = \begin{cases} \exp\left\{-\frac{\tilde{d}^2}{2\sigma_i^2}\right\} & \tilde{d} < 0 \\ 1 & 0 \le \tilde{d} \le \beta_{width} \\ \exp\left\{-\frac{(\tilde{d}-\beta_{width})^2}{2\sigma_i^2}\right\} & else \end{cases} \tag{7.3}$$

for a constant β_{width}, which describes the depth of objects, since only their
surfaces can be observed.

The mixture of Gaussians $\{ \{\mu_1, \sigma_1, \omega_1\}, \ldots, \{\mu_n, \sigma_n, \omega_n\} \}$ characterized
by their means $\mu_.$, variances $\sigma_.$ and weights $\omega_.$ is constructed with the standard

online clustering approach. This procedure is often applied because of its simplicity and speed, e.g. in background subtraction (cf. [60]). The Gaussian classified as background and the Gaussian describing the last observation are considered valid, since the current observation as well as the background, i.e. the most often used Gaussian kernels, should describe the density of space in the line of sight.

7.3.3 Measuring and Appearance Models

The tracker generates and maintains appearance models, which give a description of the target and which are searched for in consecutive frames. The observations obtained from the color cameras are modeled with color histograms (one per camera) in the proposed approach and a density model represents all 3D observations. The weighting measure for a certain configuration (sample) of the object is given by multiplication of the histogram measures and the volumetric measures. Additionally, a measure based on the distance between the previous or predicted target state and the proposed object configuration can be integrated to smooth the trajectory of the discovered target path and to prevent long jumps or large size changes. This is consistent with the fact that high velocities of the tracked objects are unlikely.

7.3.3.1 Appearance Model

In order to determine how accurate an actual configuration explains the observations of the object, each color image is projected onto the object and for every image a histogram is built up from the part of the image which coincides with the object (to do this efficiently a scanline algorithm or a bounding box of the ellipsoid can be applied). The formulae for this ray-casting approach are straightforward and are therefore omitted here.

The comparison of the histogram h_t of the target and the generated histogram h_g of a proposed target state is carried out by calculating the sum of squared differences (SSD) of all bins. The similarity measure $\delta(h_t, h_g)$ is given by

$$\delta(h_t, h_g) = \exp\left(-\frac{1}{2\sigma_{hist}^2} SSD(h_t, h_g)\right) \qquad (7.4)$$

with σ_{hist} being a smoothing constant. Other methods to measure the difference between two histograms are the histogram intersection and the Bhattacharyya coefficient. The adaptation over time of the appearance of the target h_t can be achieved with a convex combination of the histograms. This measure is evaluated separately for every image or camera respectively.

7.3.3.2 Density Model

The density model simply consists of a value describing the observed density of the object model or ellipsoid for a certain camera. The intersection density of the object model and the space model is estimated for every ToF camera by a discrete integration along each line of sight which intersects the object. For simplicity let $\{\{\mu_1, \sigma_1, \omega_1\}, \ldots, \{\mu_m, \sigma_m, \omega_m\}\}$ be the set of all valid Gaussian kernels for the pixel (x, y). Let further the line of sight through pixel (x, y) have the intersection points $X_1 = (x_1, y_1, z_1)$ and $X_2 = (x_2, y_2, z_2)$ with the ellipsoid of the object with a distance $d = \|X_1 - X_2\|$. Then the density Φ_{xy} along the line of sight through pixel (x, y) with an integration step ϵ is estimated by

$$\Phi_{xy} = \frac{1}{1 + \frac{d}{\epsilon}} \sum_{t=0,\epsilon,2\epsilon,\ldots,d} p_{xy} \left(\left\| X_1 + t\frac{(X_2 - X_1)}{d} - C_{xy} \right\| \right) \qquad (7.5)$$

with $p_{xy}(\cdot)$ being the probability of mass existing at a given point (see Eq. 7.2) and C_{xy} being the position of the camera. The density of the whole ellipsoid is computed by averaging over all densities Φ_{xy} of all pixels (x, y) which have intersection points.

A similarity measure $\delta_{vol}(\cdot, \cdot)$ between the density Φ_t of the target and the density Φ_c of the current configuration of the object is given by

$$\delta_{vol}(\Phi_t, \Phi_c) = \exp\left(-\frac{1}{2\sigma_{vol}^2} \log\left(\frac{\min\{\Phi_t, \Phi_c\}}{\max\{\Phi_t, \Phi_c\}} \right)^2 \right) \qquad (7.6)$$

with another smoothing constant σ_{vol}. Again the adaptation over time is performed by a convex combination of the densities.

7.4 Experiments

The experiments were performed with two MultiCams (see Section 3.2) and the focus lies on person-tracking, in particular head-tracking. In the first row Fig. 7.3 shows the color image and an illustration of the 3D space model, in which the body of the person was marked per hand. This ellipsoid was moved in the next frame and the photometric, the volumetric, the distance as well as all measures combined were evaluated. The results are limited to a

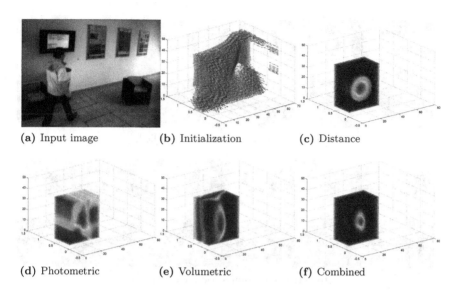

(a) Input image   (b) Initialization   (c) Distance

(d) Photometric   (e) Volumetric   (f) Combined

Figure 7.3: Characterization of the different object measures. The body of the person was marked per hand resulting in an object model and then the different measures were computed for different object centers.

selected region of the space. For this illustration an orthogonal projection is applied and therefore, the photometric measure does not change for different depth values. In the plot of the combined measures the center of the body is located with high accuracy.

In Fig. 7.4 a few frames of a head-tracking experiment using a single 2D/3D camera are displayed. Here the initialization was performed using a standard face detection method with a subsequent clustering of the 3D points of the head to remove the background. In the first row the weights of the particles were calculated with the photometric measure only. Since the head has a color similar to the frame of the LCD TV the head can easily be lost as demonstrated in the figure. In order to investigate these results further, the photometric measure was evaluated for different centers of the head and the response of the measure is plotted in Fig. 7.5. Here a position on the TV frame has an equally large likelihood than the true head position. In the second row of Fig. 7.4 both measures were applied and losing the head can then be prevented reliably. A plot showing the resulting trajectories is also shown in Fig. 7.5.

Only photometric measure applied.

Photometric and volumetric measures applied.

Figure 7.4: Experimental results for a head-tracking experiment. The head is lost when only the photometric measure is applied, but with the additional volumetric measure the target is kept.

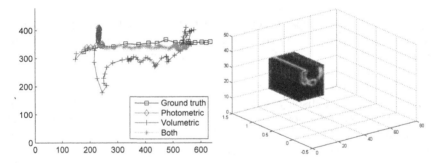

Figure 7.5: Analysis of the tracking trajectories. Left: Comparison of the resulting trajectories when applying only the photometric measure, only the volumetric measure and both measures. Right: Evaluation of the photometric measure for different object centers.

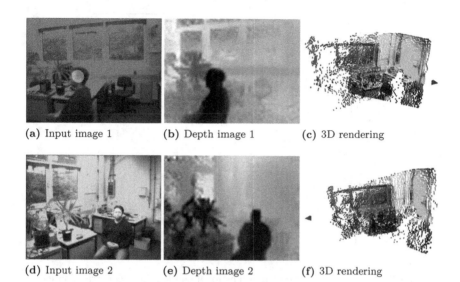

(a) Input image 1   (b) Depth image 1   (c) 3D rendering

(d) Input image 2   (e) Depth image 2   (f) 3D rendering

Figure 7.6: Images taken with two registered 2D/3D cameras and 3D renderings with textured depth measurements. The initialization is displayed in 2D and 3D.

The setup of an experiment utilizing two 2D/3D cameras is illustrated in Fig. 7.6. Here the cameras were calibrated and registered through a standard semi-automatic feature-based approach. A 3D rendering shows the positions and orientations of the cameras and the scene consists of textured depth measurements. Additionally, the initialization of the tracker is shown, which is again based on a standard head detection with subsequent clustering. The video contains fast movements relative to the frame rate and the colors are challenging. On both views the center of the head was marked per hand on all 75 frames of the video. If the (projected) center of the most likely target has an Euclidean distance smaller than 30 pixels to the true center of the head, the target is considered a match. This parameter was varied to ensure the validity of the experiment. In Fig. 7.7 the number of matches on any view using different measures and parameters are given.

Based on this experiment it can be concluded that volumetric tracking alone does not work reliably. This is quite obvious, since the tracker cannot even distinguish between the body and the head of the person. However, the volumetric measure is able to enhance the tracking accuracy significantly

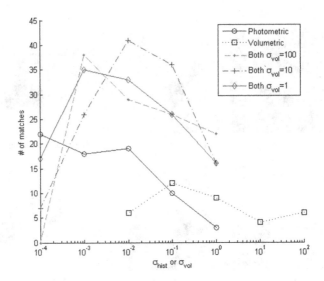

Figure 7.7: Number of correct matches for a challenging head-tracking experiment with two 2D/3D cameras. Different combinations of measures for different parameters are compared.

in conjunction with the photometric measure when compared to a tracking based on color alone.

The processing time of this tracking approach depends mainly on the number of cameras, the number of particles and the size of the target in pixels. For the head-tracking experiment with two cameras with VGA resolution and 50 particles 5 frames per second can easily be achieved on a standard office computer. Further optimization and possibly the usage of a GPU should make real-time processing possible.

7.5 Summary

In this chapter a new tracking approach was presented. It is based on the standard technique of particle filtering and combines photometric with volumetric measures to find the target. The approach is able to utilize multiple color and 3D cameras. The object is represented by an ellipsoid and the appearance of the object is modeled with color histograms. The region of a color image which originated from the object is determined by projecting the image onto the object. Additionally, the density of space is modeled with a set of Gaussian kernels for every line of sight, which makes it possible to estimate the density of space which is occupied by an object. This density is then compared to the expected density of the target in order to benefit from the 3D information.

The general approach was adopted in this work to multiple 2D/3D cameras, which combine a color with a PMD chip to obtain depth measurements. Additional color or ToF cameras can be incorporated as well without any modifications. The control mechanisms, which perform the initialization and the termination of the tracker, are relatively simple here, since this work is designed as a proof of concept and not the demonstration of a complete and working system.

In the future the implementation of more sophisticated management routines to facilitate the application of the system in activity recognition is planed. The 3D information may allow for more precise classifications of multi-agent behavior in particular. Additionally, a global object appearance model is a topic worth researching, since this might improve the tracking in non-overlapping views or the tracking of rotating objects.

8 Conclusion

8.1 Summary

The focus of this thesis lies on the extension of the area of application of PMD based depth cameras to larger scenes. Many outdoor scenes contain distances much larger than the five to ten meters which are supported by standard depth cameras. Typical devices to retrieve depth information outdoors for larger scenes are laser scanners. These measurement devices are able to provide high accuracy and precision, but are very expensive and slow, e.g. one scan per second. Therefore, novel long-range lighting setups were developed to extend the range of depth cameras, thus providing cheap and fast depth imaging while achieving reduced accuracy. One high power lighting device was designed for medium distance scenes with an opening angle appropriate for common surveillance tasks, i.e. 50 meters distance with an opening angle of 40 degrees. A complementary long-range setup for opening angles of one or two degrees and distances of up to 150 meters was demonstrated as well. Noisy depth measurements typical for larger distances affect the fusion of depth and color information severely for binocular camera setups. Thus, the ZESS MultiCam is ideal for medium- and long-range 2D/3D imaging, since its monocular setup allows for a depth independent registration of color and depth measurements.

These novel lighting devices were applied in this thesis to conduct experiments in this new area of application for depth cameras. Several processing steps are required in these applications, which exceed the demands of previous depth camera applications, e.g. a novel auto-calibration method to obtain or to validate the depth camera parameters was presented. In the area of scene observation especially background subtraction for 2D/3D videos was introduced. The proposed method is an extension of the standard approach for color videos and features a superior treatment of shadows and foreground background similarities.

Since processing of typical color and depth images is largely influence by their different resolutions, another main topic of this thesis were super-resolution methods for low resolution depth imaging. A comparison of previous methods utilizing color images in the upsampling of depth maps provides a comprehensive overview in this field. Novel methods to include the super-resolution within the main processing task were developed. A motion compensation method with joint super-resolution was introduced in order to facilitate long range depth imaging and the mobile application of depth cameras. This method utilizes the high resolution color images of the MultiCam to estimate the camera motion and thus achieves high precision. Similarly, a framework for multi-modal segmentation and super-resolution was proposed, in which both tasks influence each other in order to avoid relying on invalid depth estimates in the segmentation. Based on these experiences with the MultiCam and multi-modal imaging in general, an approach for multi-camera and mutli-modal tracking was introduced to combine different types of cameras.

8.2 Future Work

Overall, this work is meant to give a general overview of solutions, challenges and limitations for medium- and long-range 2D/3D imaging. The main focus of future work could be the increased utilization of scene geometry possibly leading to a full scene reconstruction, e.g. for the detection of obstacles or abnormalities.

Another challenging topic presents the area of image and geometry features. Here the MultiCam could again be an ideal hardware platform due to the ability to obtain registered color and depth images. Color image features a topic very actively researched and early research papers on depth image features exist. The combination of color and depth features may yield more robust and possibly a more semantically accurate features, which would allow for more accurate detection and tracking.

Lastly the combination of multiple 2D/3D cameras was only briefly explored in this work. The fusion of information collected from two or even more MultiCams could facilitate full 3D reconstruction of objects in real-time. Moreover, when employing multiple cameras for scene observation, interesting questions on how to represent the spatial data arise and promise new insights.

References

[1] B. Bartczak and R. Koch. "Dense Depth Maps from Low Resolution Time-of-Flight Depth and High Resolution Color Views". In: *Proceedings of the Int. Symposium on Advances in Visual Computing*. Las Vegas, Nevada: Springer, 2009, pp. 228–239.

[2] C. Beder, B. Bartczak, and R. Koch. "A Comparison of PMD-Cameras and Stereo-Vision for the Task of Surface Reconstruction using Patchlets". In: *Proceedings of the IEEE Conference of Computer Vision and Pattern Recognition*. 2007, pp. 1–8.

[3] J. L. Bentley. "Multidimensional binary search trees used for associative searching". In: *Communications of the ACM* 18.9 (Sept. 1975), pp. 509–517. ISSN: 00010782.

[4] K. Bernardin, T. Gehrig, and R. Stiefelhagen. "Multi-and single view multiperson tracking for smart room environments". In: *Proceedings of the international evaluation conference on Classification of events, activities and relationships*. 2006, pp. 81–92.

[5] A. Bevilacqua, L. D. Stefano, and P. Azzari. "People tracking using a time-of-flight depth sensor". In: *Proceedings of the IEEE International Conference on Video and Signal Based Surveillance*. 2006, p. 89.

[6] L. Bianchi, P. Dondi, R. Gatti, L. Lombardi, and P. Lombardi. "Evaluation of a Foreground Segmentation Algorithm for 3D Camera Sensors". In: *Image Analysis and Processing*. Ed. by P. Foggia, C. Sansone, and M. Vento. Vol. 5716. Lecture Notes in Computer Science. Springer Berlin Heidelberg, 2009, pp. 797–806. ISBN: 978-3-642-04145-7.

[7] A. Bleiweiss and M. Werman. "Fusing Time-of-Flight Depth and Color for Real-Time Segmentation and Tracking". In: *Proceedings of the DAGM Workshop on Dynamic 3D Imaging*. Berlin, Heidelberg: Springer-Verlag, 2009, pp. 58–69. ISBN: 978-3-642-03777-1.

[8]   W. Böhler, M. Bordas Vicent, and A. Marbs. "Investigating Laser Scanner Accuracy". In: *The International Archives of Photogrammetry, Remote Sensing and Spatial Information Sciences* XXXIV (2003), pp. 696–701.

[9]   D Boukerroui. "Segmentation of ultrasound images: multiresolution 2D and 3D algorithm based on global and local statistics". In: *Pattern Recognition Letters* 24.4-5 (Feb. 2003), pp. 779–790. ISSN: 01678655.

[10] D. Chan, H. Buisman, C. Theobalt, and S. Thrun. "A Noise-Aware Filter for Real-Time Depth Upsampling". In: *Proc. of ECCV Workshop on Multi-camera and Multi-modal Sensor Fusion Algorithms and Applications*. 2008.

[11] F. Chiabrando, R. Chiabrando, D. Piatti, and F. Rinaudo. "Sensors for 3D Imaging: Metric Evaluation and Calibration of a CCD/CMOS Time-of-Flight Camera". In: *Sensors* 9.12 (2009), pp. 10080–10096. ISSN: 1424-8220.

[12] O. Choi and S. Lee. "Fusion of time-of-flight and stereo for disambiguation of depth measurements". In: *Proceedings of the Asian conference on Computer Vision*. Vol. Part IV. Berlin, Heidelberg: Springer-Verlag, 2013, pp. 640–653. ISBN: 978-3-642-37446-3.

[13] O. Choi, S. Lee, and H. Lim. "Interframe consistent multifrequency phase unwrapping for time-of-flight cameras". In: *Optical Engineering* 52.5 (2013), pp. 057005–057005.

[14] D Comaniciu and P Meer. "Mean shift: A robust approach toward feature space analysis". In: *IEEE Transactions on Pattern Analysis and Machine Intelligence* 24.5 (2002), pp. 603–619.

[15] R. Crabb, C. Tracey, A. Puranik, and J. Davis. "Real-time foreground segmentation via range and color imaging". In: *Computer Vision and Pattern Recognition Workshops*. 2008, pp. 1–5.

[16] J Diebel and S Thrun. "An application of markov random fields to range sensing". In: *Proceedings of Conference on Neural Information Processing Systems*. Vol. 18. 2005, pp. 291–298.

[17] B. Drayton, D. Carnegie, and A. Dorrington. "Phase algorithms for reducing axial motion and linearity error in indirect time of flight cameras". In: *Sensors Journal, IEEE* 99 (2013). ISSN: 1530-437X.

[18] D. Droeschel, D. Holz, and S. Behnke. "Multi-frequency Phase Unwrapping for Time-of-Flight cameras". In: *IEEE International Conference on Intelligent Robots and Systems*. 2010, pp. 1463–1469.

[19] D. Droeschel, D. Holz, and S. Behnke. "Probabilistic Phase Unwrapping for Time-of-Flight Cameras". In: *Proceedings of the joint conference of the International Symposium on Robotics and the German Conference on Robotics*. Munich, Germany, 2010, pp. 318–324.

[20] D. Falie and V. Buzuloiu. "Further investigations on ToF cameras distance errors and their corrections". In: *European Conference on Circuits and Systems for Communications* (2008), pp. 197–200.

[21] M. A. Fischler and R. C. Bolles. "Random sample consensus: a paradigm for model fitting with applications to image analysis and automated cartography". In: *Commun. ACM* 24.6 (June 1981), pp. 381–395. ISSN: 0001-0782.

[22] S. Fuchs and S. May. "Calibration and registration for precise surface reconstruction with ToF cameras". In: *Proceedings of the Dynamic 3D Imaging Workshop in Conjunction with DAGM*. 2007.

[23] S. Ghobadi. "Real Time Object Recognition and Tracking". to be published. PhD thesis. Siegen, Germany: Department of Electrical Engineering and Computer Science, 2010.

[24] S. E. Ghobadi, O. E. Loepprich, F. Ahmadov, J. Bernshausen, K. Hartmann, and O. Loffeld. "Real Time Hand Based Robot Control Using 2D/3D Images". In: *Proceedings of the International Symposium on Advances in Visual Computing*. Vol. II. Berlin, Heidelberg: Springer-Verlag, 2008, pp. 307–316. ISBN: 978-3-540-89645-6.

[25] S. Gokturk and C. Tomasi. "3D head tracking based on recognition and interpolation using a time-of-flight depth sensor". In: *IEEE Computer Society Conference on Computer Vision and Pattern Recognition*. Vol. 2. 2004, pp. 211–217.

[26] D. Grest, V. Krüger, and R. Koch. "Single view motion tracking by depth and silhouette information". In: *Proceedings of the Scandinavian conference on Image analysis*. SCIA'07. Berlin, Heidelberg: Springer-Verlag, 2007, pp. 719–729. ISBN: 978-3-540-73039-2.

[27] R. I. Hartley and A. Zisserman. *Multiple View Geometry in Computer Vision*. Second. Cambridge University Press, ISBN: 0521540518, 2004.

[28] M. Harville, G. Gordon, and J. Woodfill. "Foreground Segmentation Using Adaptive Mixture Models in Color and Depth". In: *Proceedings of the IEEE Workshop on Detection and Recognition of Events in Video*. Los Alamitos, CA, USA: IEEE Computer Society, 2001, pp. 3–11. ISBN: 0-7695-1293-3.

[29] S. Hussmann, A. Hermanski, and T. Edeler. "Real-time motion su-
      pression in TOF range images". In: *Proceedings of the IEEE Instru-
      mentation and Measurement Technology Conference.* 2010, pp. 697–
      701.

[30] S. Hussmann, A. Hermanski, and T. Edeler. "Real-Time Motion Ar-
      tifact Suppression in TOF Camera Systems". In: *IEEE Transactions
      on Instrumentation and Measurement* 60.5 (2011), pp. 1682–1690.

[31] M. Isard and A. Blake. "Condensation - conditional density prop-
      agation for visual tracking". In: *International journal of computer
      vision* 29.1 (1998), pp. 5–28.

[32] F. Jager. "Contour-based segmentation and coding for depth map
      compression". In: *IEEE Visual Communications and Image Process-
      ing.* 2011, pp. 1 –4.

[33] O. Kahler, E. Rodner, and J. Denzler. "On fusion of range and inten-
      sity information using Graph-Cut for planar patch segmentation".
      In: *International Journal of Intelligent Systems Technologies and
      Applications* 5.3 (2008), pp. 365–373. ISSN: 1740-8865.

[34] T. Kahlmann, F. Remondino, and H. Ingensand. "Calibration for
      increased accuracy of the range imaging camera swissrangertm". In:
      Image Engineering and Vision Metrology 36.3 (2006), pp. 136–141.

[35] T. Kahlmann, F. Remondino, and S. Guillaume. "Range imaging
      technology: new developments and applications for people identifi-
      cation and tracking". In: *Conference on Videometrics, part of the
      IS&T/SPIE Symposium on Electronic Imaging.* Vol. 6491. Section 3.
      Citeseer, 2007.

[36] S Khan and M. Shah. "A multiview approach to tracking people in
      crowded scenes using a planar homography constraint". In: *Proceed-
      ings of the European Conference on Computer Vision.* Vol. 3954/2006.
      2006, pp. 133–146.

[37] Y. Kim, D. Chan, C. Theobalt, and S. Thrun. "Design and calibration
      of a multi-view TOF sensor fusion system". In: *Workshop on ToF-
      Camera based Computer Vision, IEEE Conference on Computer
      Vision & Pattern Recogn.* 2008, pp. 1–7. ISBN: 978-1-4244-2339-2.

[38] J. Leens, S. Piérard, O. Barnich, M. V. Droogenbroeck, and J.-M. Wagner. "Combining Color, Depth, and Motion for Video Segmentation". In: *Proceedings of the International Conference on Computer Vision Systems*. Springer-Verlag, 2009, pp. 104–113. ISBN: 978-3-642-04666-7.

[39] D. Lichti. "Self-calibration of a 3D range camera". In: *The International Archives of the Photogrammetry, Remote Sensing and Spatial Information Sciences* 37.3 (2008), pp. 927–932.

[40] M. Lindner and A. Kolb. "Compensation of Motion Artifacts for Time-of-Flight Cameras". In: *Proc. Dynamic 3D Imaging*. LNCS. Springer, 2009, pp. 16–27.

[41] M. Lindner and A. Kolb. "Lateral and depth calibration of PMD-distance sensors". In: *Lecture Notes in Computer Science* 4292 (2006), pp. 524–533.

[42] M. Lindner, M. Lambers, and A. Kolb. "Sub-pixel data fusion and edge-enhanced distance refinement for 2D/3D images". In: *International Journal of Intelligent Systems Technologies and Applications* 5.3-4 (2008), pp. 344–354.

[43] M. Lindner, I. Schiller, A. Kolb, and R. Koch. "Time-of-Flight sensor calibration for accurate range sensing". In: *Computer Vision and Image Understanding* 114 (2010), pp. 1318–1328.

[44] O. Lottner, A. Sluiter, K. Hartmann, and W. Weihs. "Movement Artefacts in Range Images of Time-of-Flight Cameras". In: *International Symposium on Signals, Circuits and Systems*. Vol. 1. 2007, pp. 1–4.

[45] O. Lottner, W. Weihs, and K. Hartmann. "Time-of-flight cameras with multiple distributed illumination units". In: *Proceedings of the conference on Signal processing, computational geometry and artificial vision*. Stevens Point, Wisconsin, USA: World Scientific, Engineering Academy, and Society (WSEAS), 2008, pp. 40–45. ISBN: 978-960-6766-95-4.

[46] O. Lottner, W. Weihs, K. Hartmann, C. for Sensor, and Systems. "Systematic Non-Linearity for Multiple Distributed Illumination Units for Time-of-Flight (PMD) Cameras". In: *Proceedings of the WSEAS International Conference on SYSTEMS*. Heraklion, Greece, 2008, pp. 752–756.

[47] O. Lottner. "Investigations of Optical 2D/3D-Imaging with Different Sensors and Illumination Configurations". PhD thesis. University of Siegen, 2011.

[48] S. H. McClure, M. J. Cree, A. A. Dorrington, and A. D. Payne. "Resolving depth-measurement ambiguity with commercially available range imaging cameras". In: *Image Processing: Machine Vision Applications*. Ed. by D. Fofi and K. S. Niel. Vol. 7538. III. San Jose, California, USA: SPIE, 2010, 75380K.

[49] OpenKinect Wiki. URL: www.openkinect.org.

[50] D. Piatti and F. Rinaudo. "SR-4000 and CamCube3.0 Time of Flight (ToF) Cameras: Tests and Comparison". In: *Remote Sensing* 4.4 (2012), pp. 1069–1089. ISSN: 2072-4292.

[51] PMD Technologies website. URL: www.pmdtec.com.

[52] T. Prasad, K Hartmann, W Weihs, S. Ghobadi, and A Sluiter. "First steps in enhancing 3D vision technique using 2D/3D sensors". In: *Computer Vision Winter Workshop*. Telc, Czech Republic, 2006, pp. 82–86.

[53] J. Radmer, P. M. Fuste, H. Schmidt, and J. Kruger. "Incident light related distance error study and calibration of the PMD-range imaging camera". In: *IEEE Computer Society Conference on Computer Vision and Pattern Recognition Workshops* (June 2008), pp. 1–6.

[54] A. Rajagopalan, A. Bhavsar, F. Wallhoff, and G. Rigoll. "Resolution Enhancement of PMD Range Maps". In: *Proceedings of the DAGM symposium on Pattern Recognition*. Vol. 5096. Munich, Germany: Springer, 2008, pp. 304–313.

[55] H. Rapp, M. Frank, F. A. Hamprecht, and B. Jahne. "A Theoretical and Experimental Investigation of the Systematic Errors and Statistical Uncertainties of Time-Of-Flight-cameras". In: *Int. J. Intell. Syst. Technol. Appl.* 5.3/4 (Nov. 2008), pp. 402–413. ISSN: 1740-8865.

[56] L. Sabeti, E. Parvizi, and Q. M. J. Wu. "Visual Tracking Using Color Cameras and Time-of-Flight Range Imaging Sensors". In: *Journal of Multimedia* 3.2 (June 2008), pp. 28–36. ISSN: 1796-2048.

[57] D. Scharstein and C. Pal. "Learning conditional random fields for stereo". In: *Proceedings of the IEEE Conference on Computer Vision and Pattern Recognition*. Minneapolis, Minnesota, 2007, pp. 1–8.

[58] I. Schiller and C. Beder. "Calibration of A PMD-Camera using a Planar Calibration Pattern Together with a Multi-camera Setup". In: *The International Archives of the Photogrammetry, Remote Sensing and Spatial Information Sciences*. Vol. XXI ISPRS Congress. 2008.

[59] S. Schuon, C. Theobalt, J. Davis, and S. Thrun. "High-quality scanning using time-of-flight depth superresolution". In: *IEEE Computer Society Conference on Computer Vision and Pattern Recognition Workshops*. 2008, pp. 1–7.

[60] C. Stauffer and W. E. L. Grimson. "Adaptive Background Mixture Models for Real-Time Tracking". In: *IEEE Computer Society Conference on Computer Vision and Pattern Recognition* 2 (1999), pp. 246–252. ISSN: 1063-6919.

[61] T. Stoyanov, A. Louloudi, H. Andreasson, and A. J. Lilienthal. "Comparative Evaluation of Range Sensor Accuracy in Indoor Environments". In: *Proceedings of the European Conference on Mobile Robots*. 2011.

[62] C. Tomasi and R. Manduchi. "Bilateral Filtering for Gray and Color Images". In: *Proceedings of the International Conference on Computer Vision*. Washington, DC, USA: IEEE Computer Society, 1998, pp. 839–846. ISBN: 81-7319-221-9.

[63] P. Vandewalle, S. Suesstrunk, and M. Vetterli. "A Frequency Domain Approach to Registration of Aliased Images with Application to Super-Resolution". In: *EURASIP Journal on Applied Signal Processing (special issue on Super-resolution)* (2006).

[64] C Wählby, I.-M. Sintorn, F Erlandsson, G Borgefors, and E Bengtsson. "Combining intensity, edge and shape information for 2D and 3D segmentation of cell nuclei in tissue sections." In: *Journal of Microscopy* 215.Pt 1 (July 2004), pp. 67–76. ISSN: 0022-2720.

[65] O. Wang, J. Finger, Q. Yang, J. Davis, and R. Yang. "Automatic Natural Video Matting with Depth". In: *Pacific Conference on Computer Graphics and Applications*. 2007, pp. 469–472.

[66] M. Wiedemann, M. Sauer, F. Driewer, and K. Schilling. "Analysis and characterization of the PMD camera for application in mobile robotics". In: *The IFAC World Congress*. Seoul, Korea, 2008, pp. 13689–13694.

[67] D. Witzner, H. Mads, S. Hansen, M. Kirschmeyer, R. Larsen, and D. Silvestre. "Cluster tracking with Time-of-Flight cameras". In: *IEEE Computer Society Conference on Computer Vision and Pattern Recognition Workshops*. Ieee, June 2008, pp. 1–6. ISBN: 978-1-4244-2339-2.

[68] Q. Yang, R. Yang, J. Davis, and D. Nister. "Spatial-Depth Super Resolution for Range Images". In: *IEEE Computer Society Conference on Computer Vision and Pattern Recognition* (2007), pp. 1–8.

Publications

[L1] B. Langmann, S. E. Ghobadi, K. Hartmann, and O. Loffeld. "Multi-Modal Background Subtraction using Gaussian Mixture Models". In: *ISPRS Technical Commission III Symposium on Photogrammetry Computer Vision and Image Analysis.* 2010, pp. 61–66.

[L2] B. Langmann, K. Hartmann, and O. Loffeld. "Comparison of Bilateral Filtering Strategies for Video Processing in Color and Depth". In: *Proceedings of the IADIS International Conference on Computer Graphics, Visualization, Computer Vision and Image Processing.* 2010, pp. 76–84.

[L3] B. Langmann, K. Hartmann, and O. Loffeld. "Depth Assisted Background Subtraction for Color Capable ToF-Cameras". In: *International Conference on Image and Video Processing and Computer Vision.* Ed. by S. P. K. Nitin Afzulpurkar Yas Alsultanny. Orlando, Florida, USA, 2010, pp. 75–82. ISBN: 978-1-60651-022-3.

[L4] B. Langmann, K. Hartmann, and O. Loffeld. "Comparison of Depth Super-Resolution Methods for 2D/3D Images". In: *International Journal of Computer Information Systems and Industrial Management Applications* 3 (2011), pp. 635–645.

[L5] B. Langmann, K. Hartmann, and O. Loffeld. "Global Multi-View Tracking utilizing Color and ToF Cameras by Combining Volumetric and Photometric Measures". In: *International Conference on Computer Vision Theory and Applications.* Vilamoura - Algarve, Portugal, 2011. ISBN: 978-989-8425-47-8.

[L6] B. Langmann, K. Hartmann, and O. Loffeld. "A Modular Framework for 2D/3D and Multi-modal Segmentation with Joint Super-Resolution". In: *Proceedings of the European Conference on Computer Vision. Workshops and Demonstrations.* Ed. by A. Fusiello, V. Murino, and R. Cucchiara. Vol. 7584. Lecture Notes in Computer Science. Springer Berlin / Heidelberg, 2012, pp. 12–21. ISBN: 978-3-642-33867-0.

[L7]   B. Langmann, K. Hartmann, and O. Loffeld. "Depth Auto-calibration for Range Cameras Based on 3D Geometry Reconstruction". In: *Advances in Visual Computing.* Ed. by G. Bebis, R. Boyle, B. Parvin, D. Koracin, C. Fowlkes, S. Wang, M.-H. Choi, S. Mantler, J. Schulze, D. Acevedo, K. Mueller, and M. Papka. Vol. 7432. Lecture Notes in Computer Science. Springer Berlin Heidelberg, 2012, pp. 756–766.

[L8]   B. Langmann, K. Hartmann, and O. Loffeld. "Depth Camera Technology Comparison and Performance Evaluation". In: *International Conference on Pattern Recognition Applications and Methods.* 2012.

[L9]   B. Langmann, K. Hartmann, and O. Loffeld. "Increasing the accuracy of Time-of-Flight cameras for machine vision applications". In: *Computers in Industry, Special Issue: 3D Imaging in Industry* 64.9 (2013), pp. 1090 –1098. ISSN: 0166-3615.

[L10] B. Langmann, K. Hartmann, and O. Loffeld. "Large Scene Reconstruction Based on ToF Cameras". In: *3D Reconstruction: Methods, Applications and Challenges.* Ed. by J. Ashworth and K. Brasher. Nova Science, 2013.

[L11] B. Langmann, K. Hartmann, and O. Loffeld. "Real-Time Image Stabilization for ToF Cameras on Mobile Platforms". In: *Time-of-Flight and Depth Imaging. Sensors, Algorithms, and Applications.* Ed. by M. Grzegorzek, C. Theobalt, R. Koch, and A. Kolb. Vol. 8200. Lecture Notes in Computer Science. Springer Berlin Heidelberg, 2013, pp. 289–301. ISBN: 978-3-642-44963-5.

[L12] B. Langmann, M. Niedermeier, H. de Meer, C. Buschmann, M. Koch, D. Pfisterer, S. Fischer, and K. Hartmann. "MOVEDETECT - Secure Detection, Localization and Classification in Wireless Sensor Networks". In: *NEW2AN.* Ed. by S. Balandin, S. D. Andreev, and Y. Koucheryavy. Vol. 8121. Lecture Notes in Computer Science. Springer, 2013, pp. 284–297. ISBN: 978-3-642-40315-6.

[L13] B. Langmann, W. Weihs, K. Hartmann, and O. Loffeld. "Development and Investigation of a Long-Range Time-of-Flight and Color Imaging System". In: *Cybernetics, IEEE Transactions on* PP.99 (2013). ISSN: 2168-2267.

[L14] O. Lottner, B. Langmann, W. Weihs, and K. Hartmann. "Scanning 2D/3D monocular camera". In: *3DTV Conference: The True Vision - Capture, Transmission and Display of 3D Video.* 2011, pp. 1–4.